A
SOLDIER'S
LAMENT

Willie Davis, Jr.

For the family of Sergeant Joseph M. Tackett, and for all who have endured loss and tragedy of any kind.

Psalm 51:10–12 (ESV)

"Create in me a clean heart, O God,
and renew a right spirit within me.
Cast me not away from your presence,
and take not your Holy Spirit from me.
Restore to me the joy of your salvation,
and uphold me with a willing spirit."

TABLE OF CONTENTS

FOREWORD
By Willie Davis Jr.

This is the book I never wanted to write.

And yet, it is the book I needed to write.

A Soldier's Lament is the story of a dream fulfilled and a life shattered. It is about what happens when the uniform you wear can no longer protect you from the weight of your own decisions. It is about the loss of a comrade, identity, peace, and the long road to healing.

There are moments in life when a single instant changes everything. For me, that moment came in Baghdad, when a negligent discharge from my weapon ended the life of Sergeant Joseph M. Tackett. Nothing in my training, my faith, or my leadership experience had prepared me for the reality that followed. In that moment, I became the thing I had feared most: the source of harm to a brother in arms.

What followed was not just a court-martial or a prison sentence. It was a spiritual reckoning.

I don't offer this book as an attempt to justify myself. There is no justification for what happened. I offer it because I believe in accountability. I believe that failure, when faced honestly, can still serve others. And I believe that redemption is real, not because I earned it, but because grace found me in my darkest place.

If you are a leader, soldier, chaplain, or someone simply struggling to make peace with your past, I pray these pages will offer more than a story. I pray they offer truth. Because while we cannot rewrite the past, we can choose how we live forward.

This is my lament, but also my offering.

- Willie Davis Jr.

PROLOGUE

Proverbs 3:5-6 ESV:

Trust in the Lord with all your heart, and do not lean on your own understanding. In all your ways acknowledge him, and he will make straight your paths.

Over the past two centuries, countless books, essays, and letters have been written about American military history, describing heroism and valor in times of war. From the American Revolutionary War in the late 18th century to the recent conflicts in Iraq and Afghanistan, numerous accounts have highlighted not only the horrors of combat but also the lessons learned from war. In every conflict in which the United States has been involved, brave men and women have shed blood in defense of their nation.

In recent history, the wars in Iraq and Afghanistan have produced many acts of heroism, continuing the proud tradition of the American military. Generation after generation, American soldiers, sailors, airmen, and Marines have shown the world what it means to be a professional, compassionate, and highly trained warrior. These individuals have taken the art of warfighting to a level unmatched in the annals of history.

I often wonder how history will explain the United States–led invasion of Iraq in 2003. How will it remember the coalition of forces that toppled Saddam Hussein's regime and sought to guide Iraq toward democracy? Promoting democracy in regions long ruled by tyranny has been a cornerstone of American foreign policy for much of the 20th century. As a young Army officer eager to serve my country, I looked forward to leading a platoon in Iraq, defending freedom, and helping dismantle a regime believed to have contributed to the 9/11 attacks and to possess weapons of mass destruction.

When I was commissioned on May 15, 2003, I hoped it would mark the beginning of a long and honorable military career, one that fulfilled my childhood dream of serving the United States with dignity and purpose. To lead soldiers is among the greatest honors a person can receive.

This memoir, however, tells a different story. It outlines one of the worst tragedies that can occur in war, one that has happened in nearly every major conflict in American history: fratricide. In my case, it was negligent fratricide, the result of my own regrettable actions as a young, commissioned officer.

For years, I have agonized over how to express my deepest apologies to the family of Sergeant Joseph Merle Tackett, whose life I took through negligence during our deployment in Operation Iraqi Freedom in 2005. Now, in

2025, I still grapple with the weight of this tragedy. Though apologies may offer some form of closure, how does one face a grieving family and say, "I'm sorry I killed your son, your husband, your brother by accident?" To me, it has never felt like enough.

It has taken me twenty years to fully come to terms with what happened—and even now, the weight of it still lingers in the quiet moments. The emotional and psychological turmoil that followed has shaped every aspect of my life. I've endured a painful spiritual awakening, and only now am I beginning to reclaim my sense of purpose. Slowly, I am finding meaning in the aftermath and trying to make amends.

Writing this memoir marks the beginning of a new purpose in my life, a step toward repentance for what happened on that fateful day in June 2005. As children, we often dream of wonderful things. But life also brings pain, tragedy, and moments we wish we could erase. Some events flip our world upside down, especially when they could have been avoided.

I've been blessed in many ways. I achieved my dream of becoming an Army officer. But my career and my dream imploded the day I was responsible for the death of Sergeant Joseph Merle Tackett. His death is an eternal wound in my heart. I choose to honor his memory by helping others in whatever ways I can. Like a spirit, he is always with me. I think of him daily, he lives in my mind and in my heart.

No matter how much I wish I could wake up from this nightmare, I know I never will. This is hard. But it must be done.

Each day, I relive the events of June 23, 2005. That day changed many lives. My negligent actions as First Lieutenant Willie Davis ended the life of SPC Tackett. How could it happen? How could I, a U.S. Army officer commissioned to lead and protect, fail so gravely in my duty?

My purpose in writing this memoir is fourfold:

1. To tell a story of personal triumph through accountability after a life-altering mistake.

2. To seek forgiveness and extend my deepest apology to those affected by the death of Sergeant Joseph M. Tackett.

3. To support others who have faced similar tragedies and demonstrate that even the greatest pain can be endured and transformed.

4. To teach future leaders, especially new officers, the importance of responsibility and leading by example.

Above all, I am a man of faith. I would be remiss

before God if I did not take steps to honor SPC Tackett's memory and use this experience to help others heal. My entire life changed due to one negligent moment with my Army-issued weapon in Baghdad. Both of us had bright futures, mine as a young field artillery officer, he as a rising NCO with the potential to become a First Sergeant or Sergeant Major.

I choose to preserve his legacy by sharing my story before, during, and after that tragic day.

To everyone who has ever endured the unexplainable: this story is for you. It is about pain, reflection, and ultimately healing. I will recount the events that led to the accident, what happened in its immediate aftermath, and how I have come to terms with it in the years since. Most importantly, I will explain how I began to move forward.

To the family and friends of Sergeant Joseph Merle Tackett, to the soldiers who served alongside him, and to all who loved him, this is my lament.

A Soldier's Lament.

May God bless all who read this. May you find inspiration, healing, and purpose in these words. And may God bless the family of Sergeant Joseph Merle Tackett and the United States of America.

"Another good mission, another good day," I thought to myself as I walked up the stairs into the Tomahawk, heading toward the lieutenants' hooch. As I moved down the long corridor, I noticed soldiers from another platoon waiting outside for their LT, Lieutenant Jones, for their nightly briefing.

"Gentlemen, what's up? How was your mission today?" I asked as I walked past.

"Nothing much, Sir," PFC Kettler responded. "How was yours?"

"Good, good," I replied, turning to acknowledge the other soldiers gathered. One by one, they greeted me: "Hi, Sir." They were soldiers from Seventh Platoon, great guys, and good troops. PFC Kettler, SPC Allen, PFC Oncale, and SPC Tackett, among them.

"Hey, Sir, how you doing?" SPC Allen asked.

"Fine, Allen," I responded.

We exchanged pleasantries, and the soldiers continued laughing and joking, something we all did to bring a sense of normalcy to deployment. They even joked about me, which I didn't mind. A little levity goes a long way during war.

As the jokes continued, I suddenly realized my weapon wasn't positioned safely. It wasn't angled downward at a 45-degree angle, and worse, the safety selector was set to "semi" rather than "safe." As I flipped the switch, I noticed my weapon was unintentionally pointed toward the soldiers and then, suddenly...

POW .... "Oh my God ...Oh my God," I said in horror. "Tackett...Tackett." I screamed as my fellow soldiers scattered away from the one deadly shot that was fired from my weapon. "What have I done?" I thought to myself as SPC Tackett fell to the ground, I saw blood spew from his cranium.

I ran towards him, straddled him, and placed my hands on his head to stop him from bleeding so profusely. As I saw, his eyes began to roll up into his head. I knew what had happened! I knew what I had done. "I killed him! My God, I killed him. No! This can't be happening! Tackett, please—stay with me! No! No! God, please, no!

PART I: THE CALLING AND THE CLIMB

1
A DREAM OF SERVING

1 Peter 4:9–11 (ESV):

"Show hospitality to one another without grumbling. As each has received a gift, use it to serve one another, as good stewards of God's varied grace: whoever speaks, as one who speaks oracles of God; whoever serves, as one who serves by the strength that God supplies, in order that in everything God may be glorified through Jesus Christ. To him belong glory and dominion forever and ever. Amen."

Many people dream, but how many actually achieve their dreams?

When I was young, I dreamed big. I believe children's dreams are often rooted in fantasy, but as they grow, those dreams begin to blur the line between fantasy and reality. And when they finally come true, those dreams become real in every sense.

Everyone, at some point in childhood, is asked, "What do you want to be when you grow up?" Many say, a firefighter, a professional athlete, or simply someone rich beyond imagination. But those weren't my dreams.

All I ever wanted was to serve my country in the Army of the United States of America. That desire goes back as far as I can remember.

I first began thinking seriously about serving in the military in 1991. At the time, the United States and coalition forces were preparing to launch Operation Desert Storm, a mission to expel Iraqi forces from Kuwait following Iraq's illegal occupation and annexation of the country the previous year.

"Daddy… Daddy?" I asked emphatically.

"Yeah," he replied, walking into my room with his hands in his pockets, glancing at the television.

"Are we at war? Why are we at war, and who are we at war with?" I demanded.

As my father looked at the television screen, he turned to me and said, "We're at war with a tyrant named Saddam Hussein and the country of Iraq, a country in the Middle East that invaded its neighbor. Now I guess we must go after them."

"Oh my God, Daddy!" I said with alarm. "They could invade us too. We live on the coast!" I began to pack my most prized possessions, my keyboard, boom box, and Nintendo. Afraid that we would be next.

My father chuckled as he left the room, saying, "Boy, you're crazy. This is the United States of America. That's not going to happen."

As it turns out, he was right.

On the eve of the coalition's launch of Operation Desert Storm, I sat with my father watching the news. NBC anchor Tom Brokaw broke down the strategy and scope of the military operation. From that moment on, I was in awe of the United States military machine.

I started fantasizing about becoming a soldier or a U.S. Marine. As I grew older, I began studying military history, especially that of the United States. I learned about and admired American military leaders, from the Revolutionary War to modern conflicts, men like General Joshua Chamberlain, General Ulysses S. Grant, General Benjamin O. Davis Sr., and General Colin Powell.

While my friends dreamed of becoming professional athletes, I dreamed of serving my country as a military officer.

Everything I did as a child was, in some way, preparing me for that goal. I studied hard, stayed out of trouble, and threw myself into sports with determination. I immersed myself in military history and the study of great leaders, trying to understand what made them exceptional. Serving my country wasn't just a dream, it was my deepest ambition, the fire that shaped my choices and defined who I wanted to become.

As a first-year student in high school, I joined the Boy Scouts of America, Troop 8 in East Orange, New Jersey. I was a latecomer—most scouts join in elementary school—but I didn't care. I loved every moment of it. Through the Scouts, I learned valuable leadership skills and had my first meaningful encounters with the structure and discipline that mirrored military life. I worked hard, embraced every challenge, and eventually earned the rank of Eagle Scout—a milestone that filled me with pride and prepared me for the journey ahead.

One of the most influential experiences came in the spring of 1996. Our troop took a sightseeing trip to Baltimore, Maryland, and Washington, D.C. We visited the Smithsonian, the Capitol, and the National Air and Space Museum. But the defining moment for me was our visit to the United States Naval Academy in Annapolis, Maryland.

Before that visit, I didn't even know the military had academic institutions dedicated to training and molding future officers. At the Academy, I got a glimpse of what it meant to be a commissioned officer in the Navy or Marine Corps. Inspired, I began researching the other service academies: West Point and the Air Force Academy. I learned about their requirements, high GPAs, strong SAT/ACT scores, leadership experience, and a congressional nomination.

It was then that reality set in. By the tenth grade, I realized I was academically behind for that level of competition. I was a good student with a 3.30 GPA and a member of the National Honor Society, but I hadn't been adequately challenged at Orange High School in New Jersey. I'd taken several unnecessary courses at the behest of the State of New Jersey in my first two years, which kept me from enrolling in more rigorous subjects.

By senior year, I knew I needed to be strategic. I looked for colleges that offered two things: a strong international studies program and an Army ROTC option. I wanted to test whether military life, especially the Army, was really for me.

My choice wasn't popular among many friends and family, especially my mother. Still, I made a bold decision. After receiving an informational packet in the mail, I chose a two-year military junior college: the New Mexico Military Institute (NMMI) in Roswell. Known simply as "The Institute" to students and alumni, it was also a prep school for service academies. It offered a high-quality education in a disciplined, military-structured environment. That's exactly what I wanted.

My time at NMMI was transformative. I learned

what it meant to be a leader, mentored by figures like COL Jack R. Fox and LTC Antonio Pino (Ret.). There was nothing else I wanted more than to be a commissioned officer in the U.S. Army.

I enrolled in Army ROTC and learned about military structure, Army values, and the standards expected of officers. For me, it wasn't just about the four-year active and four-year reserve commitment. This was where I belonged. I wanted to be a strong, ethical leader, someone who could make sound decisions under pressure. I wanted to make the Army my career.

My experience at NMMI only solidified that desire. It paid off when I was awarded a two-year, campus-based Army ROTC scholarship to Morehouse College, a historically Black college in Atlanta, Georgia.

God, I was excited!

2

PUT ON THE FULL ARMOR OF GOD

Ephesians 6-11:

"Put on the full armor of God, so that you can take your stand against the devil's schemes. For our struggle is not against flesh and blood, but against the rulers, against the authorities, against the powers of this dark world and against the spiritual forces of evil in the heavenly realms. Therefore, put on the full armor of God, so that when the day of evil comes, you may be able to stand your ground, and after you have done everything, to stand. Stand firm then, with the belt of truth buckled around your waist, with the breastplate of righteousness in place, and with your feet fitted with the readiness that comes from the gospel of peace. In addition to all this, take up the shield of faith, with which you can extinguish all the flaming arrows of the evil one. Take the helmet of salvation and the sword of the Spirit, which is the word of God."

Operation Iraqi Freedom III.

Two months after the United States and coalition forces invaded Iraq and began Operation Iraqi Freedom, I was commissioned as a 2nd Lt. in the United States Army on

15 May 2003 along with my friend's, I was ready to accept the responsibility of leading men and women and helping further the cause of democracy and freedom and upholding American traditions and values wherever my country sends me. I took an oath to defend the United States Constitution and to lead the soldiers who were under my command to victory under the direction of the President of the United States. In my mind, outside of my faith in God and in love, there is no greater duty than the service of one's country. I was now a commissioned officer in the United States Army. And my duty to my country was more important than ever.

I received my orders about a month ago. I was to report to the Field Artillery Officer Basic Course No Later than 16 September 2003 at Fort Sill, Oklahoma. I was so excited. I had received the job that I wanted and the duty station I wanted, Schofield Barracks, Hawaii, home of the Army's 25th Infantry Division. I requested that I stay in Atlanta as a Gold-Bar recruiter, mainly so I can stay in Atlanta and be able to spend time with a certain woman who began to catch my eye in the early part of 2003. She was my best friend's cousin, and I thought she was such a remarkable woman. She happened to be in Atlanta studying for her degree in Divinity. Her name was Kristen, and she was something else! During my time at Morehouse, Kristen

.

would look after my things every summer and we talked every so often. We were both always busy. Never really spoke to each other for more than a few months at a time. In the early spring of 2003, I began to take more notice of her. She was smart, kind, and grounded, and I came to see qualities in her that I hadn't appreciated before.

I have always held tightly to a few things I love: my faith in God, my country—which I was proud to serve in the United States Army—and, yes, the Green Bay Packers. At that chapter of my life, I also loved Kristen and was grateful for the time we shared together.

She did not have my love for the military as I do. Nobody in my inner circle or immediate family did, for that matter. It is not every day that a young Black man from the inner city grows up with the love of the military and of the country the way that I do. Most African Americans of previous generations are very skeptical of America and of the military for various reasons. But not me, people say that I am a different breed, and that was alright. I married Kristen at the dawn of 2004, while I was still enrolled in the Field Artillery Officer Basic Course at Fort Sill, Oklahoma. And I was very much in love. I had decided to switch my dream assignment with another Lieutenant in my class from the 25th Infantry Division in Hawaii, to the 3rd Infantry

Division stationed at Fort Stewart, Georgia, so I could be close to her, even though it was three and a half hours from Atlanta. Even though I knew that I was deploying to Iraq within the coming year, I had to be close to her. I carried two residences, one in Atlanta and one in Hinesville, Georgia, the city adjacent to Fort Stewart, to make it work.

When I arrived at Fort Stewart, Georgia, it was early March 2004, and I had graduated from the Field Artillery Officer Basic Course before I went on leave to spend time with my wife, Kristen, three weeks before. Man, I was both excited and nervous. I was at the precipice of my Army career, and I knew that my next four years as a lieutenant were to be the most important of any officer's career. These are the years in which I will be molded as a leader of soldiers and learn my craft. My biggest prayer was to be placed in a unit with fine officer leadership and even better non-commissioned officers. Because, as a Lieutenant, I understood that it was the enlisted soldiers in which an officer in the United States Army learns his or her job.

I was initially assigned to the 1st Battalion, 41st Field Artillery Regiment. I had heard great things about the Battalion Commander there, LTC Patrick Antonelli, through word of mouth around Fort Stewart. Other soldiers had informed me that he was a good battalion commander with superior knowledge of fire direction. I had also heard that he was tough on his officers physically and professionally.

In my initial meeting with LTC Antonelli, he welcomed me to the battalion and told me that he was glad to have me. But he informed me that my time with the battalion would be a short one. "Sir?" I said. "Why will my time in the battalion be short?" "Lt. Davis," he stated emphatically. "The 3rd Infantry Division is reorganizing into four brigades, and the two field artillery battalions will be instructed to deactivate their Charlie batteries so that they can be reactivated in a new battalion within the forthcoming 4th Brigade, 3rd Infantry Division." Colonel Antonetti paused for a minute as he seemed to be looking at his battalions' officer billets. "Lt. Davis, there are no openings in my Alpha or my Bravo firing batteries. However, Charlie Battery is currently in need of a Fire Direction Officer. Therefore, you will be assigned to Charlie Battery, 1st Battalion, 41st Field Artillery Regiment, Charlie Dawgs." "CPT. Alric Francis is your Battery Commander." "If I recall correctly, he was a tall African American man whose command presence was impossible to ignore." I thought silently to myself. "Well, Lt. Davis, off to Charlie Battery you go," he said. "The battalion S1 will escort you there. Good luck Lt. Davis. CPT Francis is a good commander; you will be in good hands there. You are dismissed, welcome to the 1st Battalion, 41st Field Artillery Regiment and the 3rd Infantry Division," he said. "Thank you, Sir," I replied. I rendered him a salute, did an about face, and exited his office.

The Battalion S4, a young CPT, escorted me to the Charlie Battery office and introduced me to CPT Ric Francis. CPT Francis was a senior Captain, maybe a couple of years or so before he was to pin on major at the time when I arrived at the unit. He had been in the Army for ten years and had seen five deployments between the 1st Cavalry Division out of Fort Hood, Texas, and the "Marne" Division. So, he was a very seasoned officer, and I very much liked him and his leadership style. CPT Francis was a tall, bald African American man who believed in simple leadership by example, or at least that was what I would say to describe his leadership style. He was the type of officer who did what his soldiers did and damn sure was not afraid to train them if they were not up to speed. Especially when it came to his young Lieutenants!

In my first meeting with him as his new young lieutenant, he explained to me what his philosophy was and what I was going to do." Lt Davis, welcome to Charlie Battery 1-41, and I am glad to have you...so tell me about yourself, Lt." So, for the next hour or so, I began to tell CPT Francis about myself and my goals in relation to the service. After we got through the introductions and the preliminaries, he began to talk about the mission at hand. "Well, Lt. Davis, we will be deploying to Iraq in support of Operation Iraqi Freedom late this year or early 2005," he said with conviction. I knew I was going to deploy in at least a year or so before I gave up my dream assignment to the 25th Infantry Division in Schofield Barracks, Hawaii. "You will be my new FDO, Fire Direction Officer for 2nd Platoon FDC, Fire Direction Center," he continued. "But we have a lot of work to do before then. We are going to deactivate Charlie Battery 1-41 in a couple of months and stand-up Alpha Battery, 1st Battalion, 76 Field Artillery Regiment, and the new 4th Brigade, 3rd Infantry Division." I thought to myself for a moment as CPT. Francis turned his head and looked towards his computer. I realized that in the next few months, there was going to be a lot of work. "You will also replace LT. Avila as Battery Supply Officer, and your Supply Sergeant is SGT. Crosby." "Remember, Lt., know that being a 2nd lieutenant means that this is your time to learn your craft and how to lead soldiers. Soak up all the knowledge that you can from your non-commissioned officers, your fire direction chief, and your soldiers," he said. We have some good soldiers, some good men in this battery, as he shall soon see."

Over the course of the next ten months or so, I began to follow the advice of CPT Francis, which was to learn from my non-commissioned officers, fellow lieutenants, and work on my personal physical fitness so that I could endure the rigorous desert environment of Iraq. The battery leadership, in my opinion, was superb at the time CPT Francis was the Battery Commander. I know that most soldiers think that their units are the best in the Army in every way, but my units' leadership was top notch!

During that time, CPT Francis challenged me and my fellow lieutenants to hone our military skills, physical fitness, and to get to know our soldiers and NCOs' strengths and weaknesses. I was the youngest lieutenant in the battery when I first arrived. I was not only young in experience, but I was also young in age. I was twenty-two years old, the youngest of all the lieutenants and just out of the Officer Basic Course and very green. But at least I had experienced non-commissioned officers and officers to learn from.

The senior lieutenant in the battery and the one whom I had come to both look up to and respect was 1st Lt. Russell Porter. Lt. Porter was a really tall guy, about twenty-five or twenty-six years of age. He was the Charlie Battery, Executive Officer and 2nd Platoon Leader. The first time I met Lt. Porter, I thought that he was not a very approachable person, but that was just at first glance, I had soon learned the opposite. Cpt. Francis first introduced me to Lt. Porter when I first walked into the Battery before morning physical. Lt. Porter was the first one I had met. He was coming out of the 2nd platoon leaders' office. "LT Porter, this is Lt. Willie Davis, your new fire direction officer." CPT Chen said. Lt. Porter extended his hand towards me. "Hey man, how are you doing? The name is Russ." "How are you doing, Russ? I said as I extended my hand towards him. "Nice to meet you, Russ," he replied. "Outside of CPT Francis, Lt. Porter was the only other officer in the battery who had combat experience, serving with the 3rd Infantry Division's 2nd Battalion, 7th Infantry Regiment as a Fire Support Officer. As I looked past Lt. Porter, I saw another soldier who looked as if he was coming towards me to introduce himself. "Hey Willie, my name is Ryan, Ryan Avila, so you are replacing me as Russ's FDO." "I guess so," I replied. Lt. Avila was the complete opposite of Lt. Porter in regard to his height. He was short in stature, and I came to find out that he was serious but loved having fun and he loved his soldiers. "Well, I believe that you are inheriting a great group of soldiers. Take care of them for me."

As the day continued, I met SSG Burt who was to be

my fire direction NCO, my platoon sergeant, SFC Robinson and my platoon gunnery Sergeant, SSG Polynice. These were the men with whom I was going to train and who I was going to learn from in order to become a better officer and leader. These soldiers and officers would be both my mentors and soldiers in arms for the next couple of years, and I was very much looking forward to seeing what we would accomplish as a unit.

We all talked for a few minutes, and Lt. Porter introduced me to some of the lower enlisted soldiers and NCOs before I heard the Battery First Sergeant call for PT formation outside the Charlie Battery Headquarters. "Platoon Sergeants, form up outside for PT formation," commanded 1st Sergeant Garnet Taylor. As I looked across the battery conference room, I saw a 1st Sergeant Taylor who had to be a six-foot-four, two-hundred-and-sixty-pound intimidating figure of a man, who approached me and introduced himself. "How are you doing, Lt. Davis? I am 1st Sgt Taylor, glad to have you with us, Sir." "Thank you, 1st Sgt, glad to be here," I replied. "Are you ready to run, Sir? The BC wants to do a Battery Run today." Get ready, Sir, get ready, BC don't play when he runs,' he stated as he ran out of the Battery headquarters. "Get ready, Charlie Dawgs, get ready, baby, form up," he ordered as I followed the other officers out to formation. I thought to myself that this is going to be real fun! I am going to really like it here.

For the first couple of weeks, I shadowed LT. Avila, he showed me the ins and outs of my job as the 2nd Platoon Fire Direction Officer. "Willie, as far as the soldiers are concerned, you have a very good group of soldiers, led by SSG Burt, who is young compared to the other NCOs, but he is very smart and knows fire direction." Lt Avila said. "He has taught me a thing or two, so listen and learn from him. He is a good NCO who will not steer you wrong." Lt. Avila ran down the weekly schedule, which consisted mostly of maintenance of vehicles and soldiers' equipment, and (DST) Digital Sustainment Training as well as Physical Fitness Training. "The most important thing that you need to do as 2nd Lt. is listen and learn from the NCOs in this Battery, we have some good ones and we have some not so good ones, in which I am sure you will be able to distinguish between the two," commented Lt. Avila.

So, for the first four months of my young military career, I learned what it meant to be a fire direction officer in a mechanized heavy artillery battery with the exceptional arsenal of the M109A6 Paladin artillery firing system in today's military. CPT Francis was always challenging me and always correcting me as he gave me room to make my own mistakes and to quickly learn from them.

Man, I felt he was always on my back, always riding me more than the other LTs. Even though I had become frustrated at times, I knew that all he was doing was preparing me and making me better for the challenges ahead.

While at Officer Basic Course (OBC), before I had made the decision to switch my assigned unit with a member in my class so that I could be closer to my wife, I had learned that the 3rd Infantry Division was to redeploy to Iraq in support of Operation Iraqi Freedom III (OIF III) in early 2005. As LTC Antonetti told me when I first came to 1-41 FA, Charlie Battery was to be deactivated and reactivated as Alpha Battery, 1st Battalion, 76th Field Artillery Regiment, a part of the new 4th Brigade, 3rd Infantry Division. Therefore, in a matter of seven months, we had to stand up a new battalion that consisted of a new table of organization and prepare for the ensuing year-long deployment in Iraq. CPT. Francis instructed his officers and non-commissioned officers to prepare their soldiers for some tough and long days ahead. "Gentlemen," CPT Francis stated as he addressed his senior leadership during a battery meeting. "I have met our new battalion Commander...LTC Daniel Pinnell and I really liked it. "In a lot of ways, he was just like me," CPT Francis stated. "He believes in the same things I do and under him, I believe we will be in good hands in Iraq." He continued. I couldn't wait to meet my new battalion commander. I told myself that if he was even half the leader and mentor that CPT Francis had been, then the battalion would be in good hands.

On 21 June 2004, the Army deactivated Charlie Battery, 1st Battalion, 41st Field Artillery and activated Alpha Battery, 1st Battalion, 76 Field Artillery. No more three-gun Paladin platoons and three-battery field artillery firing battalion. And on the 24th of June 2004, after 17 years and 4 months of being inactive, the 1st Battalion, 76th Field Artillery Regiment was reactivated. The battalion was to be referred to as the "Patriot" Battalion in reference to 1776.

When all the other battalions were activated or reassigned, the army completed the reorganization of the 3rd Infantry Division and activated the new 4th Brigade "Vanguard", with Colonel Edward C. Cordon and Engineer Officer serving as the brigade's first commander. It was the beginning of a hectic six months, which was the time allotted to us to prepare and train before we were to deploy to Iraq. The unit transition was completed. Now we were preparing ourselves and our soldiers for a successful year-long deployment in Iraq.

3
SENT FORTH IN STRENGTH
Joshua 1:9:

"Have I not commanded you? Be strong and courageous. Do not be afraid; do not be discouraged, for the Lord your God will be with you wherever you go."

It was in the sweltering Georgia heat, during one of those long, demanding summer afternoons, that I first met Lieutenant Colonel Daniel A. Pinnell, a man who would forever redefine my understanding of military leadership. At that time, I was immersed in the physically exhausting task of helping relocate the Alpha Battery supply room, sweat-soaked, focused, and completely unaware that my first encounter with one of the most influential leaders in my military journey was just moments away.

"Hey Willie," said LT Alberto Reynoso, one of my fellow lieutenants and the first platoon Fire Direction Officer. His voice carried a hint of urgency and reverence. "Our new battalion commander is on the premises. He is coming this way."

Immediately, my heart rate ticked up a few notches. The news jolted me into action, and I dropped what I was doing without hesitation. Dusting off my uniform, I quickly stepped out of the supply room into the battery's common area, the air thick with anticipation. The room was brightly lit, but the energy changed the moment the door opened and in walked a tall, bald Caucasian man, exuding a kind of silent authority that commanded immediate respect. He was accompanied by Captain Curry, Battalion S-4.

With a confident stride and an approachable smile, the new battalion commander addressed us.

"Gentlemen, how are you doing?" he asked in a calm, assured tone, extending his hand to each of us.

"We're doing fine, Sir," we replied instinctively and in unison.

"Lieutenant Davis, Sir, 2nd Platoon Fire Direction Officer," I said, stepping forward and introducing myself.

He looked me square in the eye and asked, "So, any words of wisdom for me?"

The question caught me slightly off guard. It was not the kind of question one expected from a battalion commander meeting a junior officer for the first time. Yet, it spoke volumes about his leadership style, open, disarming, and intentional. That brief interaction marked the beginning of what would become one of the most formative relationships of my military career.

The months that followed were a whirlwind. Our

battalion's training schedule was relentless. Every square on the calendar was filled with field exercises, tactical drills, planning meetings, and preparation tasks. We were laser-focused on a single objective: combat readiness. In July, we conducted a grueling battalion-wide field training exercise, followed closely by the division's comprehensive "Marne Focus" operation in early August. Then, in October, came the ultimate test, our final pre-deployment rotation at the Joint Readiness Training Center (JRTC) at Fort Polk, Louisiana.

This was not standard preparation. It was a crucible.

Although we were field artillerymen by trade, trained to shoot, move, and communicate with speed and precision, our upcoming mission demanded something different entirely. This time, we would not just be launching shells from Paladins or calculating fire missions from behind secure lines. We were being asked to take on roles traditionally performed by infantry, maneuvering in convoys, escorting high-value personnel, and engaging threats directly. For many, it was unfamiliar territory, requiring a shift not just in tactics, but in mindset.

Our battalion was tasked with forming twelve-gun truck platoons from our 400-man, all-male combat arms unit. Once in theater, we were to receive forty-eight up-armored

HMMWVs and serve as mobile security details for U.S. Embassy staff and Iraqi dignitaries. These convoys would take us across Iraq, ferrying political and administrative leaders to meetings with ministries and coalition officials. We would help lay the groundwork for Iraq's nascent democracy, ensuring these crucial interactions occurred safely in one of the world's most dangerous environments.

For many of our senior NCOs, battle-hardened field artillerymen, this was a new frontier. While some of the junior soldiers had been exposed to infantry tactics during initial entry training, our sergeants and staff sergeants had spent the bulk of their careers focused on artillery-specific skills. Now, they were brushing up on infantry doctrine, clearing buildings, conducting route reconnaissance, and mastering convoy operations. It was a daunting transition, but our leaders were determined to rise to the occasion.

Captain Francis, our battery commander, was the anchor during this turbulent period. His tone during our officer and senior NCO meetings was always serious, deliberate, and imbued with a sense of urgency.

"Gentlemen," he said during one such meeting, "this deployment will not resemble the last one. We are not sure of our exact tasking yet, but platoon leaders are unlikely to be commanding Paladin crews. You are going to be leading convoys on the roads, through towns, through danger. There are a multitude of potential missions, and we need to be ready for all of them."

Then, with a pause and a firm voice, he added: "Make no mistake. We will be well-trained."

Our training intensified. Under LT Porter's guidance, 2nd Platoon began studying FM 7-8: Infantry Rifle Platoon and Squad, absorbing knowledge with the intensity of students cramming for final exams. We rehearsed tactical movements, learned dismounted patrol procedures, and ran drill, after drill after drill. At the same time, we did not abandon our artillery fundamentals, we continued practicing our core competencies until they were second nature. We knew our mission could change at any moment, and we had to be ready for anything.

When we arrived at the Joint Readiness Training Center (JRTC) in October, just four months into my role as a Fire Direction Officer, I was given new orders: I would take over as a Platoon Leader. It was a defining moment. Captain Francis, First Sergeant Taylor, and LT Porter reorganized the battery into seven platoons, and I was assigned to lead one of them, with Staff Sergeant Johnny Patterson serving as my platoon sergeant.

Our platoon configuration was straightforward, yet

formidable: twelve soldiers, including myself, a platoon sergeant, four drivers, four hatch gunners, and two truck commanders. Together, we would form one of the many gun truck platoons designated to support Operation Iraqi Freedom III.

To my immense relief and gratitude, the soldiers and NCOs assigned to my platoon were some of the finest I could have hoped for. Their backgrounds were diverse, their talents many, and their years of experience ranged from five to nearly two decades. Each of them brought something unique to the table.

My platoon sergeant, Staff Sergeant Johnny Patterson, was someone I already knew. He was a quiet professional, calm, reserved, and thoughtful. Initially, I had reservations about his level of drive. Unlike some of the other senior NCOs who made their presence known through vocal motivation and visible intensity, Patterson carried himself with subtlety. But I quickly came to appreciate that behind his quiet demeanor was a reservoir of experience and wisdom.

"Wassup, LT?" He would say with a smile whenever we crossed paths. It was his signature greeting, always laid-back but genuine.

As we settled into our roles, I asked Patterson about

the other NCOs in our platoon, SSG Golf, SGT Clifton Cogdell, and SGT Frederick Taylor.

"Out of all three, Sir," Patterson said thoughtfully, "I am most impressed with Sergeant Cogdell. He just got back from a fourteen-month deployment in Iraq with the 3rd Armored Cavalry Regiment out of Fort Sill."

"That must be tough, coming back just to get ready to go again," I said.

"It is, Sir. He is leaving his family behind at Fort Sill while he deploys with us. But make no mistake, his experience will be a huge asset."

Patterson was right. Cogdell was sharp, dependable, and knowledgeable. He had seen the fight up close and understood what was at stake.

As December 2004 arrived, deployment became more than a concept, it became a reality. We transitioned from woodland camouflage BDUs to the tan-and-brown Desert Combat Uniforms (DCUs). We turned in old gear and drew new, preparing our kits for the long year ahead. There was a sense of finality in the air.

Our soldiers were also preparing their families for what would be an extended absence. For some, it meant updating wills and arranging finances. For others, it meant difficult conversations with spouses and children. For me, it meant emotionally preparing my wife, who was still in Atlanta finishing her degree. We had already been living apart, but this was different. This was war.

Inside, I was a storm of emotions, excitement, fear, pride, and anxiety. Would I lead my soldiers well? Would I make the right decisions under fire? Would I be able to bring everyone home?

I did not agree with the political rationale for the war, but my private opinions had no bearing on my duty. I had taken an oath to support and defend the Constitution, to obey orders, to lead soldiers. That oath carried weight, and I bore it with reverence.

LTC Pinnell, always candid, addressed our battalion during a late December 2004 formation.

"As much as I'd love to bring every soldier home safely, I must be honest," he said. "We are going to war, and we must prepare for the reality that not everyone may come back. That is the harsh truth. Our strength will be in how we prepare, how we fight, and how we lead."

I stood there, still and reflective, praying that no one from my platoon would become one of those names etched in granite or remembered in silence.

We were granted three weeks of leave before deploying. My wife and I traveled to New Jersey to spend time with our families, and for me, that meant seeing my mother.

My mother had always been a rock in my life. Strong, assertive, and a person of faith, she never wanted me to join the military. Two of her brothers had served, and when I first shared my dreams of becoming an Army officer, she cried. I was her youngest, her "Little Willie." She worried about me with every fiber of her being. It was exhausting at times, but she was my mom, and I understood.

Now, as I prepared for my first combat deployment, her fears only deepened.

"You be careful over there," she said, holding my face in her hands. "Watch what you eat. Do not be doing any of that... friendly fire."

"Ma, why would you say that?" I asked, surprised.

"Because anything can happen in war," she replied. "Just do your job well. Stay prayed up. And come home."

Her embrace that day wrapped around me like a suit of armor—strong, unwavering, and fiercely protective. It was as if, in that single moment, she was shielding me from the weight of the world, holding everything broken within me together with nothing more than the strength of her love. Before returning to Fort Stewart, I made time to visit the people who had molded my life: my father, my Aunt Shirley, my mentor Coach Horace Hooper, and my childhood best friend Jason, who was more like a brother to me. Every Christmas Eve, Jason and I would play chess and celebrate with his family, his sister Tara, his cousin Kristen, and others who had become like family to me. That night was a tradition, a sacred space of laughter, competition, and brotherhood.

Coach Horace Hooper, Jason's father, had helped shape me into the man I was becoming. He had helped train me for the Army Physical Fitness Test, challenged me to think critically, and taught me the importance of persistence.

"Willie," he told me, "If you don't get it right the first time, get your ass up and try again."

That night, surrounded by people who loved and believed in me, I felt grounded. Supported. Ready.

As I hugged each of them goodbye, I hoped my soldiers were doing the same, cherishing their loved ones, holding onto those last moments of peace before we all stepped into the unknown.

Because very soon… we were going to war.

4

COMPLACENCIES COST

Isaiah 41:10:

"So do not fear, for I am with you; do not be dismayed, for I am your God. I will strengthen you and help you; I will uphold you with my righteous right hand."

The moment finally came when I deployed to Iraq in support of Operation Iraqi Freedom III. It was February of 2005, and I still remember the solemn weight I felt stepping onto that aircraft at Hunter's Army Airfield in Savannah, GA, knowing that we were heading into the very heart of a war-torn land where uncertainty reigned and danger was constant. I was part of the advance party crew for the battalion, which meant I would be one of the first boots on the ground.

Our battalion, the 1st Battalion, 76th Field Artillery Regiment, known as the "Patriots," had recently been reactivated and assigned to the 4th Brigade Combat Team, 3rd Infantry Division at Fort Stewart, Georgia, on June 24, 2004. In just under seven months, we would find ourselves in the epicenter of the conflict, Baghdad, Iraq. It felt like everything had been set in motion so quickly, yet with great precision, as if destiny had pulled us into a moment in history that would shape each of us forever.

I have always been the kind of leader who wanted to know what I was walking into. I was not comfortable with relying on assumptions or theories from textbooks when it came to the lives of the men under my command. I wanted to observe firsthand, to learn from those who had already endured the rigors of deployment. That was one of the reasons I was glad to be part of the advance party. I wanted to study the (TTPs), Tactics, Techniques, and Procedures of the 1st Cavalry Division, which had been operating in-theater for more than a year and understood the lay of the land in a way that no training scenario back home could replicate. So, I decided to go out on missions in order to be better prepared for what was to come.

Those first days and weeks in Iraq were disorienting but also clarifying. There was no room for indecision or hesitation. Being part of the advance party allowed me to settle in quickly, gather valuable intelligence, build relationships with our predecessors, and mentally prepare for the year-long mission ahead. I began embedding with the 1st Battalion, 82nd Field Artillery Regiment, joining their convoys and patrols, absorbing every detail of their in-country operations. These missions gave me the insight and operational awareness I would need to lead effectively once the rest of my unit arrived.

Our mission was complex and, in many ways, abstract. At least, that is how it seemed on the surface. Officially, we were there to help the Iraqi people rebuild and reestablish their government, to support them as they attempted to stand on their own two feet after years of dictatorship, war, and upheaval. But what that really meant in the day-to-day was something far more challenging and far more personal. It meant navigating shifting loyalties, a fractured infrastructure, and a population trying to survive amidst the crossfire of history.

We operated in a world where the enemy didn't wear a uniform. There were no front lines, no clear boundaries. Just streets and alleyways where any vehicle could be rigged to explode, and any civilian might be watching—reporting our every move. In this chaotic and volatile landscape, one name loomed large over everything we did: **Al-Qaeda in Iraq**.

Al-Qaeda in Iraq, or AQI, was more than just a terrorist network. It was an ideology, a shadowy force that fed off the instability we were fighting to control. Led at the time by Abu Musab al-Zarqawi, AQI aimed not only to drive out American forces but to fracture Iraq from within. Their strategy wasn't conventional warfare, it was terror. Bombings, executions, and the deliberate incitement of sectarian violence between Sunni and Shia Iraqis were their weapons of choice.

Our battalion provided security in support of the U.S. Embassy, a mission that required discipline, vigilance, and an unwavering sense of purpose. We weren't just fighting an enemy; we were trying to build something amid the wreckage. Security. Stability. A future. For my platoon, this meant every convoy had to be carefully planned and executed with discipline and vigilance. We weren't just avoiding IEDs—we were rambling through the middle of a sectarian powder keg, and AQI was holding the match. Little did I know at that time that, over the course of that deployment, the Patriots battalion conducted an average of 18 daily combat missions across the entire Multi-National Forces–Iraq area of responsibility. We tallied 3,849 combat patrols, traversed over 530,138 miles of treacherous terrain, and managed to safely deliver 61,183 diplomats, contractors, and senior military officers to their destinations.

I was entrusted with leading 4th Platoon, Alpha Battery, a four-gun track platoon composed of 12 of the finest soldiers I have ever known. Each man brought his own strength, his own story, his own reason for being there. And together, we formed a cohesive unit, a brotherhood bound not just by uniform and mission, but by the understanding that our lives depended on each other. There was no room for ego. No space for selfishness. Every decision I made as a leader, every command I issued, carried real consequences. These were not theoretical exercises anymore. This was war. And yet, amid all the pressure, I found something deeply meaningful in the responsibility I held.

There were moments, quiet ones, often in the stillness after a mission or in the hum of an idling convoy, when I would reflect on the surreal nature of it all. I would look around at the sunbaked buildings, the wary stares of local Iraqis, the endless desert sky stretching over a land older than time itself, and I would think: This is history, and I'm living in it. We all are.

But even more than that, I would think about my men. Their safety. Their families. The burden of leadership weighed heavily on me, not just in the tactical sense, but morally and spiritually. I knew that whatever happened out here, I would carry it with me for the rest of my life. I wanted to make sure that I did right by them, that I gave them every chance to survive.

The rest of the battalion arrived in-country at the beginning of March 2005, and it marked a pivotal turning point in our operational readiness. Their arrival infused renewed energy into our efforts, and we immediately began the right-seat/left-seat ride process with the 1st Battalion, 82nd Field Artillery, a necessary transition ritual where outgoing and incoming units worked together to ensure a seamless handoff. For us, it meant shadowing their operations, learning from their hard-earned experiences, and preparing to take full ownership of our mission supporting the U.S. Embassy in Baghdad for the next year.

It was during this transition period that I truly began to establish a routine for myself and for 4th Platoon, Alpha Battery, 1st Battalion, 76th Field Artillery Regiment. Routine, in a place like Baghdad, was a strange and often fleeting concept, but it was crucial. It provided a sense of structure in a chaotic environment. It gave my men something to rely on, something familiar, even as every mission brought new unknowns. I made it my priority to ensure that the platoon operated with the highest standards of safety and discipline, whether we were out on the roads or back on the Forward Operating Base (FOB). In a theater of war where a single mistake could mean disaster, vigilance was everything.

Our operational tempo was relentless. We ran missions at least five days a week, often more, navigating some of the most dangerous routes in and around Baghdad. Our primary responsibility was to transport diplomats, contractors, and military leaders to various government facilities, infrastructure sites, and key installations throughout the capital. These weren't simply escort missions; they were strategic movements that often took us deep into volatile zones where IEDs, ambushes, and insurgent activity were not only common, but they were also expected.

I can still remember the tension in the air during our missions to places like the Iraqi Ministry of Electricity, located just outside the Green Zone. That facility was critical to rebuilding Iraq's infrastructure, but its proximity to contested areas made it a hotbed of potential conflict. Or the Al-Dora Oil Refinery, far to the south in Baghdad, an essential piece of Iraq's economic engine and a frequent target of insurgent interest. Each destination brought its own set of logistical, tactical, and emotional challenges, and no two days were ever the same.

Despite the repetition of mission briefs and route checks, nothing ever felt routine. The city itself was always shifting, allegiances changed, tensions flared, intel evolved, and we had to adapt constantly. There were days when I felt like we were walking a tightrope with no safety net beneath us, relying solely on our training, instincts, and trust in one another to get through.

Because of this, I was relentless when it came to safety and procedural discipline. Every mission, every movement, every radio check, and convoy formation mattered. I made it my duty to drill those principles into the hearts and minds of my men, not out of fear, but out of necessity. I needed them to understand the gravity of our actions, the thin line between success and tragedy that we walked each day. Whether it was a pre-mission equipment check, ensuring proper muzzle awareness, enforcing rules of engagement, or conducting after-action reviews, I treated each step as sacred. The margin for error was zero, and the cost of complacency was too high to gamble with.

But safety didn't end on the roads. Back on the FOB, I remained equally committed to enforcing structure, discipline, and accountability. The FOB was our only sanctuary, if you could call it that, and I viewed it as my responsibility to ensure that my soldiers had a place to rest, reset, and remain mission focused. I encouraged them to stay mentally sharp, physically strong, and spiritually grounded, even as the days wore on and the fatigue of deployment began to sink deeper into our bones.

Looking back, those early months in Iraq taught me a great deal, not just about combat leadership, but about human endurance, adaptability, and the quiet courage that exists in the hearts of ordinary soldiers doing extraordinary things. We were more than just a platoon of artillerymen; we had become a unit of protectors and ambassadors, navigating a foreign land on behalf of our country and the values we swore to defend.

However, as time went on, the cracks in my leadership began to slowly reveal themselves. What started as small lapses became glaring shortcomings that I could no longer ignore. One of the most critical failures was my inconsistency in enforcing the very standards I preached, especially regarding weapon clearing procedures and muzzle awareness. These weren't just check-the-box safety protocols; they were lifesaving habits, ingrained into every soldier from day one. But over time, amid the exhaustion, the routine, and the psychological wear of operating in a combat zone, I began to falter, not only in holding my men accountable but in holding myself accountable as well.

At first, it seemed minor. A rushed clearing procedure here, a weapon slung carelessly there. But in hindsight, those moments were warnings, red flags of complacency that I should have addressed with urgency. My responsibility as a leader was not just to complete the mission, but to ensure that my platoon returned home alive and whole. And yet, I allowed the erosion of discipline to go unchecked. I had become so focused on the demands of the mission, so consumed by the day-to-day grind, that I failed to enforce the very foundations of soldiering.

And that failure... that lapse in judgment... would ultimately change the course of my life and cost another man his.

The morning of 23 June 2005 began like so many others in Iraq: I rose before first light, pulled on my Desert Combat Uniform, and made my way to the DFAC for a sparse breakfast: hard-boiled eggs, stale toast, and powdered coffee. My mind was already focused on the day's mission. We were heading to the Al Dora Oil Refinery, deep in southern Baghdad. That meant punching out of the Green Zone and charging down Route Irish, one of the most dangerous roads in the country, notorious for IEDs and ambushes. I had studied its twists and choke points, briefed my platoon on its perils, and yet each time my heart still tightened at the thought of running it.

The mission went off without a hitch, our convoy cleared the Al-Dora Oil Refinery site and headed back through the entry control point of FOB Honor. As we idled by the clearing barrels, I dismounted and handed my rifle to my driver, assuming he would run the standard clearing drill. In my fatigue, I skipped the last vital steps: no chamber check with my finger, no visual inspection. Confident that everyone else had cleared their weapons, I turned and joined the platoon as we rolled toward the motor pool. We began the familiar shutdown procedures, toggling ignition switches, removing helmets, stowing gear, and preparing to secure our vehicles for the evening. In that split second of complacency, I traded vigilance for convenience, unwittingly setting the stage for a tragedy none of us would ever forget.

Later that evening, I stopped by the DFAC for dinner and carried my tray along the usual path back to my quarters. I paused when I saw the entire 7th Platoon seated outside their barracks, just as they did every night. As I approached, they greeted me with the same warmth and humor that defined Alpha Battery, 1-76 FA. We exchanged jokes and pleasantries. I cherished these moments of camaraderie, many of these soldiers were my age or only slightly younger, tied to me by duty and friendship.

We joked. One of the soldiers, though I never caught his name, teased me about my "AFLAC duck" nod and quack. I laughed heartily, but then, as I looked down at my weapon, a cold knot formed in my stomach: my rifle's safety selector was flipped to "Semi" instead of "Safe." In my fatigue, I'd slipped into the bad habit of toggling the switch with my thumb and then walking away without double-checking.

Realizing the danger, I proceeded to correct it. My rifle was nearly parallel to the deck, and my trigger finger was resting in the well. In the blink of an eye, that careless moment ended with a thunderous crack, one reckless twist that would change everything.

POW .... "Oh my God ...Oh my God," I said in horror. "Tackett...Tackett." I screamed as my fellow soldiers scattered away from the one deadly shot that was fired from my weapon. "What have I done?" I thought to myself as SPC Tackett fell to the ground, I saw blood spew from his cranium.

I ran towards him, straddled him, and placed my hands on his head in an effort to stop him from bleeding so profusely. And as I saw, his eyes began to roll up into his head. I knew what had happened! I knew what I had done. "I killed him! My God, I killed him. No! This can't be. Tackett, please stay with me! No, don't die! No! No!

PART II:
THE FALL

5

NIGHTMARE IN THE TOMAHAWK: ONE ROUND, ENDLESS CONSEQUENCE

Psalm 38:4:
"My guilt has overwhelmed me like a burden too heavy to bear."

After that moment, everything that followed unfolded in a blur, as if the world around me had shattered and time had fractured into chaos. I remember collapsing under the weight of disbelief, my knees buckling beneath me as I cried out in anguish. I screamed for Sergeant Tackett to come back, pleading, bargaining, yelling to a heaven that stayed silent. The words spilled from my mouth, raw and frantic:

"Please! Don't die! God, please, don't let him die!"

Two soldiers rushed to my side, dragging me away from the scene. My legs refused to cooperate, and I stumbled as they guided me into another room in the Tomahawk. They sat me down gently on a bench, their arms steadying me, one of them whispering, "It's okay, Sir. Just breathe." But how could I breathe? My lungs were tight with panic. My chest felt like it had caved in. There was no oxygen, only the crushing, suffocating weight of guilt and disbelief.

That's when I saw him, Chaplain Kim. He arrived swiftly, his face drawn, his movements purposeful. Without hesitation, he sat beside me and placed a firm yet calming hand on my shoulder. I couldn't contain the emotion that surged up inside me any longer. I turned to my platoon sergeant and choked out the words:

"I killed him. I killed Tackett."

My voice was barely more than a whisper, yet those words rang louder than any explosion I had ever heard. The pain in them was unmistakable shame, disbelief, and heartbreak bound into a single breath.

My platoon sergeant froze, visibly shaken. "No, Sir," he said softly, the tremble in his voice betraying his own grief, "You didn't kill him. It was an accident. You didn't mean to." But even as he said it, I could see the disbelief swimming behind his eyes. He wanted to protect and comfort me. But deep down, we both knew something irreversible had happened.

Chaplain Kim leaned closer and asked quietly, "Are you okay?"

I looked up at him, tears streaming down my face, my voice breaking apart like shattered glass:

"I killed him, Sir. I killed him." I said it over and

over, as if repetition might somehow relieve the burden or make it less true.

"It was an accident… but I killed him."

I could hear those words echoing inside my own head, louder than any gunfire I had ever heard in combat. They weren't just words; they were a verdict. My soul had spoken, and there was no taking it back.

The moment froze around me. My sense of self, my understanding of leadership, my future, it all collapsed under the weight of that terrible realization. The sorrow I felt in that instant was unlike anything I had ever known. And it would never fully release me.

As more soldiers began to gather, I saw the expressions on their faces: shock, disbelief, confusion. Eyes wide, mouths slightly agape, some unable to speak, others whispering prayers. The weight of the moment had begun to settle over all of us like a fog, thick and suffocating. A wave of sorrow and disgust came crashing over me, and I felt myself unraveling. My cries turned to hysteria, my thoughts fragmented. I was no longer grounded in reality.

Medics arrived. Their hands were firm but kind as they assessed me, trying to bring order to the chaos within me. I remember flinching as they approached, confused and bewildered, and then the last thing I saw before the world went black was Lieutenant Aliva, watching me silently as I was loaded into the back of a HMMWV. I turned to him, barely able to form the words:

"I'm sorry... I'm so sorry..." And then everything faded.

The next thing I remember was waking up in a hospital bed later that night. The sterile white walls. It all felt surreal. CPT Francis and 1st Sgt Taylor stood at my bedside, their expressions somber as they came to check on me.

For a brief moment, I thought I was waking from a nightmare, a terrible, vivid dream that would vanish as soon as I opened my eyes. But as the haze of sedation began to lift, the truth settled in with crushing weight. This wasn't a dream.

This wasn't the familiar cement-block walls of my room in the Tomahawk. It was a medical facility, quiet, clinical, and cold. And what had happened wasn't a figment of my imagination. It was real. All of it.

The sorrow I had tried to escape found me the moment I regained consciousness. No one had to say a word. I knew. And I wished more than anything that I didn't.

As I sat up in the hospital bed to greet my commander, tears began to stream down my face, slow at first, then uncontrollable. The weight of what had happened bore down on me again, but this time, my thoughts turned to someone I hadn't yet considered: Sergeant Tackett's family.

"Sir," I said, my voice trembling, "was he married? Did… did he have any children?"

It was the first time the thought had crossed my mind, and the moment it did, my heart sank. I winced at the possibility, as if it were a blade being slowly pressed into my chest. The idea that children might now be fatherless because of me, that a wife might be left alone, shattered by grief, was almost too much to bear. My sorrow deepened into something darker, more suffocating. I could hardly breathe.

CPT Francis placed a steady hand on my shoulder and answered gently, "He was married… but he didn't have any children."

His words landed like a stone in my chest, relief, and anguish, all tangled into one. I nodded slowly, the tears still falling, wishing with every fiber of my being that I could undo what had been done.

As I continued to weep, CPT Francis leaned in closer. His voice was calm but heavy with gravity.

"Lieutenant," he said, "there are going to be some tough times ahead. Prepare yourself."

He paused, then repeated it, slower this time, as if to ensure the words would sink in.

"There are going to be some tough times ahead. Keep your head up."

I looked up at him, eyes red and swollen, and he met my gaze without flinching. There was no judgment in his expression, only a stern compassion that said more than words ever could. Then, without another word, he turned and began to walk out of the room.

1st Sergeant Taylor stepped forward next, echoing the same sentiment. He placed a firm, grounding hand on my shoulder.

"Sir," he said quietly, "there are going to be some tough times ahead."

He gave me a solemn nod, then followed CPT Francis out, the sound of their boots fading as they returned to the FOB.

As they left, I lay back down in the hospital bed, my mind spinning with anguish. What could possibly be tougher than this? I thought. This was a complete nightmare; one I couldn't wake up from. Sergeant Tackett was gone. Gone. And there was nothing I could do to change that. The weight of it crushed me. It was all my fault, I kept repeating in my head, over and over. I turned my face into the pillow, haunted by the events of the evening, the sounds, the faces, the terrible finality of it all. Eventually, exhausted and broken, I drifted back into sleep, though rest never truly came.

Later that night, I was awakened by an unexpected visitor, a man I had met just two weeks earlier: Major General William Webster, the commanding general of the 3rd Infantry Division. So much had changed in those two weeks. What had once been a routine introduction now felt like a lifetime ago. The atmosphere was heavy, and his presence carried the weight of both authority and compassion.

As he approached my bedside, I instinctively began to stand in order to greet him, but he quickly motioned for me to remain seated. I obeyed, my posture tense with shame. The guilt pressed heavily on my chest, and before he could say a word, I spoke through trembling lips.

"Sir... I don't deserve the 'Outstanding Soldier' coin you gave me," I said, my voice breaking. "Not after what happened."

Just two weeks earlier, he had presented that coin to me and my platoon in recognition of our combat readiness during his inspection of our mission preparations. It had been a proud moment, an affirmation of our training, discipline, and leadership. But now, that memory felt distant and undeserved, overshadowed by the tragedy that had unfolded.

"I failed, Sir," I added through sobs, my eyes cast

downward. "I let everyone down."

General Webster stepped closer, his voice firm but compassionate. "No… no. This was an accident."

He reached out and gently took my hand, clasping it between his own. The weight of my guilt spilled over.

"I'm so sorry, Sir," I muttered, my voice cracking. "It was my fault."

He said nothing at first, perhaps there were no words that could ease the anguish I felt. Instead, he continued to pat my hand softly, trying in his own quiet way to steady me as I broke down completely. The grief surged through me, raw and unstoppable, and I began to cry hysterically.

Unable to offer more comfort and visibly affected himself, General Webster gave a respectful nod and quietly excused himself from the room, leaving me alone in the crushing silence of my sorrow. It was my fault; it was all my fault.

6

INTERNAL RECKONING

Hebrews 4:12:

"For the word of God is alive and active. Sharper than any double-edged sword, it penetrates even to dividing soul and spirit, joints, and marrow; it judges the thoughts and attitudes of the heart."

After spending the night at the Combat Support Hospital, I was transferred to the U.S. Embassy where I was placed on suicide watch. Those next days, however many they were, blurred together into a haze of fluorescent lights, locked doors, and overwhelming silence. The experience of being watched around the clock, of not even being trusted to be alone with my own thoughts, put me in a dark place mentally.

Alone in that room, my mind spiraled through a storm of questions, each one louder than the last. How are my soldiers doing? Are they okay? Did they make it through the night? Are they angry? Do they blame me?

But the question that echoed more than anything else, the one that clung to me like smoke, was this:

"How in the hell was my weapon loaded?"

This suicide watch experience was unlike anything I'd ever known.

Never in my life had I felt so completely alone, so hollowed out by guilt, so angry at myself, and so utterly unwilling to go on living. In those long, endless hours under observation, I kept reliving the night of June 23, 2005. Over and over. Each time the memory returned, it cut deeper.

Looking back now, I understand why they put me on suicide watch. It was the right decision. Because in those moments, I was slipping, I was beginning to entertain the darkest of thoughts.

There were times I was called out of that cold, isolating room, mostly to eat, sometimes to speak with LTC Breitenbach, my first therapist. She was a calm and steady presence amid a storm I couldn't begin to navigate on my own. A woman from Indiana. She brought with her a patient heart, a listening ear, and a quiet strength that gently pierced through the fog of my despair. In those earliest conversations, she didn't try to fix me. She simply sat with me, asked questions that mattered, and listened without judgment. Through her guidance, I began what I now understand to be the slow, painful journey toward self-understanding, and perhaps, in time, forgiveness.

It wasn't about healing everything overnight. It wasn't even about healing at all, not yet. It wasn't about forgetting either. It was simply about surviving, one hour at a time, then the next. And in that dark, silent place, those small victories became lifelines. They were the thin threads holding me back from the edge, keeping me from going off the cliff that loomed ever so close.

And still, even as I clung to those fragile wins, the question gnawed at me: How was my weapon loaded? I thought it was cleared. I believed it was safe. But belief doesn't save lives, actions do. I eventually had to face the painful truth: proper clearing procedures weren't followed. I hadn't verified the chamber. I hadn't taken the steps I drilled into my soldiers. The burden of that failure lay squarely on my shoulders. As a platoon leader, the standard I walked past became the standard I accepted. And in doing so, I became an example of what not to be.

These were the kinds of conversations I had with Colonel Breitenbach. Quiet, intense, sometimes halting. She would ask gently, "What happened?" And I would try to piece it together, not just the sequence of events, but the why. Why had I trusted the process instead of verifying it? Why did my discipline slip in that moment? Why did my nervous habit of flicking the safety switch back and forth go unchecked?

In those early sessions, she didn't try to excuse anything. She didn't let me hide behind rationalizations. But she also didn't let me drown in guilt. She asked the tough questions, but she did it with compassion. And through that process, I slowly began to confront the truth, not just the external facts, but the internal reckoning that had to come and would continue for years, even decades to come.

After spending a week in the medical wing of the U.S. Embassy under suicide watch, I was formally transferred to Headquarters and Headquarters Battery (HHB), 4th Brigade Troops Battalion, 4th Brigade, 3rd Infantry Division. My new destination was FOB Prosperity; a secure base nestled within the heavily fortified Green Zone of Baghdad. The relocation marked a significant shift, not just geographically, but emotionally and mentally. I was leaving behind the familiar terrain of FOB Honor and stepping into a new, more uncertain phase of my life.

FOB Prosperity was where I was to await the outcome of my impending General Court-Martial, a process that loomed over me with the weight of final judgment. Every moment was tinged with anxiety, dread, and reflection. I knew that the coming weeks would shape the course of the rest of my life.

While at FOB Prosperity, I was temporarily assigned to the Brigade Force Protection Unit, specifically the Explosive Ordnance Disposal (EOD) team. My role, though limited, was to assist in their critical mission of locating and neutralizing Improvised Explosive Devices (IEDs) and Explosively Formed Penetrators (EFPs), lethal threats hidden beneath roadsides or embedded into walls, waiting to strike.

It was dangerous work, and I was still under intense emotional strain. But being around soldiers again, supporting a mission, even in a diminished capacity, gave me a temporary sense of purpose. At least for a few hours each day, I could focus on something other than the tragedy I had caused. Yet even amid the tension of roadside threats and high-stakes patrols, nothing could distract me from the trial ahead or from the deep ache that followed me everywhere I went.

It was during this time that I had the opportunity to meet and speak with my brigade commander, Colonel Edward Cardon. He came to see me within a day of my reassignment to the 4th Brigade Troops Battalion, and I'll never forget that meeting.

Colonel Cardon was a respected officer and an exceptional leader of soldiers. Commissioned as an Engineer Officer from the United States Military Academy in 1982, he was one of the first in his field to command an infantry brigade, a testament to his capabilities and the trust the Army placed in him. What struck me most about Colonel Cardon was his calm command presence and his ability to truly listen. He wasn't just someone who spoke with authority, he listened with intention and with compassion.

I didn't know what to expect from him. Part of me feared the worst, that he would berate me, break me down further, or express the disappointment I already felt so heavily within myself. But much like Major General Webster before him, Colonel Cardon responded with understanding, empathy, and a quiet strength.

When I met him, the first thing I did was offer a heartfelt apology.

"Sir," I said, choking back emotion, "I am deeply sorry for the actions that led to Specialist Tackett's death. I was derelict in my duty as an officer. No words can ever justify what happened."

Colonel Cardon looked at me—not with condemnation, nor with the cold detachment I had feared—but with the solemn, steady gaze of a leader who understood the crushing weight of responsibility. There was no judgment in his eyes, only a quiet acknowledgment of the burden I carried. He didn't rush me or cut me off; he let me speak, allowing my words to stumble out, raw and unfiltered. In his presence, I didn't feel like a failure to be discarded or a case study in poor leadership. For the first time since the incident, I felt something I thought I had lost: my humanity. I was broken, yes, shattered in ways I could barely articulate—but somehow, in that moment, I was still a man, still a soldier, still human in his eyes.

During my time at FOB Prosperity, most of my days were consumed with mission objectives as part of the Force Protection Team, all while I waited for updates and meetings with my defense counsel. It was a strange and unsettling season of my life. I felt disoriented—like a shadow of myself—existing in a war zone without a weapon, the only soldier walking around FOB Prosperity unarmed.

At first, I expected to be confronted, corrected, or questioned. I braced myself for someone to demand, "Where's your weapon, Lieutenant?" But to my surprise, no one ever did. Not a single person asked. And so, I walked the perimeter and common areas of the FOB aimlessly, alone with my thoughts and my guilt, still trying to process the gravity of what had happened.

I felt completely isolated. Though no one said it aloud, I knew most soldiers had heard the whispers. They knew what had happened. They knew what I had done, or at least, what they thought I had done.

One afternoon, while having lunch in the dining

facility, I met another officer, someone who, for once, wasn't afraid to speak to me. She didn't flinch, didn't avoid eye contact, didn't seem to carry the quiet judgment I had grown used to seeing in others. She was an African American woman, and I later learned her name was Lieutenant Shannel Watson.

As we talked, I discovered she was from Newark, New Jersey—not far from where I grew up. There was an unexpected comfort in that small connection, a sense of familiarity I hadn't felt in weeks. The more we spoke over lunch in the DFAC, the more I began to feel like myself again—even if only a little. Then, almost as if by divine coincidence, we realized we had a mutual friend: someone I had gone to high school with, and she had played college basketball with.

That simple thread of connection reminded me that even in the midst of isolation, shame, and fear, I wasn't completely alone. Through the conversations we began to have in that dining facility, I came to learn more about Lieutenant Watson—not just as an officer, but as a person.

She was a devoted soldier, a loving wife, and, above all, a woman of deep faith. It didn't take long to notice how often she spoke about how blessed she was, even in the middle of a war zone. Her strong faith in God wasn't something she merely talked about; it radiated through her actions, her words, and the way she carried herself. There was calm strength in her demeanor, a quiet confidence that seemed unshaken by the chaos around us.

For me, those moments with her were like brief glimpses of light piercing through the heavy clouds of guilt and despair. She never pried into my situation or treated me differently because of it. Instead, she offered the simple gift of conversation—a reminder that I was still seen, still human, still worth engaging with.

One day, as we sat across from each other in the dining facility eating lunch, I decided to open up to her. For days, the words had been lodged in my throat, but now they pushed their way to the surface.

"Shanell... do you know why I'm here? Why am I at FOB Prosperity?" I asked, my voice barely above a whisper.

She set her fork down and looked at me with calm, steady eyes. Before she even answered, I felt a tightening in my stomach. My palms grew damp, and my chest felt heavy. She was the first person I had chosen to be this vulnerable with, in this war-torn environment, and it scared me. Yet, something about her made it easier. She had an approachable spirit and the rare gift of being a good listener, someone who could sit in silence and still make you feel heard.

"Yes, Davis, I do," she said softly. "I know why you're here… and I feel for you."

The lump in my throat swelled as I forced out the words, I had been too ashamed to speak.

"It was my fault," I said, my voice cracking. "I didn't follow proper clearing procedures, and it resulted in the death of one of my soldiers. I—I am ashamed."

For a moment, there was only silence between us, the kind that feels sacred rather than awkward. Shanell reached across the table slightly, her eyes never leaving mine.

"God will help you through this, Davis," she said gently but firmly. "Be strong and courageous. I don't know everything you're going through, but I do know this—you are loved. And I can see the pain you're carrying inside. You don't have to carry it alone."

Her words didn't erase the guilt, but they pierced through the darkness I had been suffocating in. It was as if, in that moment, God Himself had sent me a reminder— through her—that I wasn't beyond redemption.

Within a couple of weeks of arriving at FOB Prosperity, I was assigned a JAG officer to serve as my defense counsel, as it had become increasingly clear that all roads were leading to a General Court-Martial. Her name was CPT Fansu Ku. CPT Ku was a petite woman of Asian descent, sharp, methodical, and exceptionally intelligent.

She had a calm, calculated demeanor, but there was a warmth about her, a quiet grace that made it easier to speak candidly in what was, without question, the darkest chapter of my life.

I still remember our first meeting with vivid clarity. I sat alone in the waiting area of the Judge Advocate General's Office, my nerves raw as I anxiously scanned the walls lined with documents and case summaries. One in particular caught my attention, a case involving another officer charged with negligent homicide. But it wasn't just a single charge. There were three counts of negligent homicide listed.

I froze. For a moment, it felt as if I were staring into a mirror, seeing my own reflection in that paper. My heart began to pound, each beat echoing louder in my chest as the weight of my situation pressed in on me.

"Not one… but three?" I whispered under my breath, my mind racing. What if I had killed three soldiers? The thought hit me like a punch to the gut. I couldn't imagine carrying that kind of weight. My chest tightened as I sat there, silently weeping—not just for that officer, but for the soldiers whose lives had been lost.

As the tears threatened to spill, I found myself wondering, what would happen to him? And then, almost immediately, the question turned inward: What will happen to me?

Just then, CPT Ku emerged from her office. Her presence was unassuming, but her energy was composed and confident. She offered me a polite smile and extended her hand.

"Lieutenant Davis, how are you?" she asked.

I replied, "I'm doing okay, Ma'am," though the truth was far more complicated than that.

She introduced herself formally as Captain Fansu Ku, my appointed defense counsel, and invited me into her office. Her voice was calm, deliberate, and reassuring, despite the seriousness of what we were about to discuss.

As we sat down, CPT Ku wasted no time addressing the gravity of my situation. Her demeanor remained steady and professional, but I could sense the seriousness behind her calm tone. She opened a legal pad and, with deliberate clarity, said:

"Lieutenant Davis, I need you to explain to me exactly what happened and how it happened."

I had expected this moment. I knew it was coming. And yet, hearing those words made my chest tighten all over again. I had replayed the events of that night over and over in my mind for days, each time hoping the details would make more sense, or that something would change. But they never did.

I rose slowly from my seat, unable to remain still under the weight of her request. With my hands clenched, voice trembling slightly, I began to recount everything, detail by detail. I described how the day had unfolded, the convoy to the Al Dora Oil Refinery, our return to FOB Honor, the clearing barrels, the moment of laughter, the mispositioned safety selector switch, the sound of the shot, the panic, the horror, and the heartbreak. I poured it out with painstaking care, as if each word had the power to bring clarity or at least provide some measure of truth.

She listened quietly, never interrupting, her eyes steady and focused. When I finished, the silence that followed felt deafening. She nodded slowly and leaned forward with a serious expression on her face.

"Given the circumstances," she began carefully, "I strongly recommend you submit paperwork to resign your commission in lieu of a General Court-Martial."

The words struck me like a blow to the chest.

She continued, "This route would avoid a full court-martial trial. However, it would result in an Other Than Honorable Discharge from the United States Army."

It was a sobering recommendation, but I understood.

This wasn't about avoiding accountability. It was about navigating the military justice system in the most realistic and responsible way possible, given the facts of the case and the consequences that loomed over me. Still, hearing those words, "resign your commission," made everything feel so final, so irreversible. A career I had dreamed of and worked toward for years was now hanging by a thread, and this path felt like an admission that it was already lost.

I looked at CPT Ku, her expression unreadable yet calm, and asked her the question that had immediately surfaced in my mind:

"Do you think the Commanding General will approve it?"

She paused for a moment, weighing her words carefully before replying.

"I don't think so," she said frankly, "but it's worth a shot."

Her honesty was unsettling, but I appreciated it. There were no false reassurances, just the plain, hard truth of the situation. Even so, "worth a shot" was better than no shot at all.

I thought that this was the Army's version of mercy. A version of mercy, quite frankly, I did not deserve.

It felt like the final nail. That recommendation,

logical, perhaps even merciful under the circumstances, hit me with the force of a thousand pounds. This was no longer theoretical. The Army career I had built, the uniform I had once worn with pride, was slipping away piece by piece. Still, I knew she was doing her duty to prepare me for every possibility, even those that broke my heart.

Until the date of my trial, I continued traveling to Camp Victory twice a week to meet with CPT Ku. Each visit was a solemn reminder of the weight of what lay ahead. Every conversation we had slowly chipped away at any lingering disbelief or denial, leaving in their place a cold, unforgiving truth: the life I once knew, the career, the purpose, and the dream, was over.

What brought me some small measure of comfort during this period was the presence of an old friend from college, 1st LT Danielle Smith, someone I had received my commission alongside. I remember the first time I saw her after we landed in-country. At first, I wasn't sure. She had on her desert camouflage patrol cap and a pair of dark sunglasses, and I thought she might've been staring at me. I squinted, trying to make her out through the bustle of the landing zone, and then it hit me -*That's Smith.*

Without thinking, I moved toward her, almost tripping over a tent tie-down in the process. We both laughed when I caught myself. It was such a small moment, but it felt like a burst of familiarity in a world that had become so foreign. She was assigned to a postal battalion that provided logistical support to U.S. forces throughout the theater, and her base wasn't far from mine. After one especially difficult visit with CPT Ku, when the weight of everything I was carrying became too much, I realized I needed someone familiar. I reached out to her.

Danielle's familiar face was like a lighthouse in the storm, steady, reassuring, and unmistakably real. When I saw her in front of her battalion headquarters, she greeted me with the same warm, level gaze she had always carried back in our ROTC days. For a moment, it felt as though time had folded in on itself, as if the years hadn't passed, as if war hadn't weathered us, and everything we'd endured since commissioning had left no trace. It was a flicker of something I hadn't felt in a long time: safety.

There was no awkwardness in her demeanor, no fumbling for words, no trace of pity, just presence. That rare kind of presence that asks nothing and expects nothing, the kind that anchors you in the middle of emotional chaos. The kind that doesn't need to say, *I see you,* because everything in her eyes already had.

After some small talk, just enough to remember what

normal sounded like, I found myself speaking more freely than I had in weeks. I told her what had happened. About the night that changed everything. About the accidental discharge. About SPC Tackett. I didn't sugarcoat it. I couldn't. The words felt foreign leaving my mouth, like they belonged to someone else. But Danielle never flinched.

She listened. She didn't try to fix it. She didn't offer hollow reassurances. She just sat there, nodding gently, letting me speak. At one point, I paused, waiting for her expression to shift into judgment or discomfort. It didn't. Her eyes held mine, steady, clear. "I'm so sorry," she said quietly. "I can't imagine what you're carrying. But I'm here. You're not alone in this."

That moment cracked something open in me. I hadn't realized just how isolated I'd become, how far removed from basic human kindness. In war, you build walls to survive. But Danielle, with a few words and a familiar presence, began to tear one of mine down.

A few days later, during another visit, the hammer fell hard. I was handed the Army's official response to my resignation request. One word sealed my fate: Denied. Below it, stamped without hesitation, were the formal charges: Article 119, Involuntary Manslaughter. Two additional counts of dereliction of duty. Each line felt like a sentence in a language I never wanted to learn. My hands trembled. My breath caught.

There it was, in black and white, undeniable, and unmerciful. Any illusion that I might quietly walk away from this, that the Army might show grace, was crushed beneath the full weight of military justice. The system I had once served now stood in judgment over me. And in that moment, I felt the ground beneath my identity shift. No more hiding. No more bargaining. They had spoken. And now, I would answer.

I remember telling Danielle, unsure of how to form the words. "They're pressing charges," I said, staring at the floor.

She didn't speak right away. Instead, she leaned forward, placed her hand on mine, and looked at me with the same calm strength she'd always carried. "I figured they would," she said gently. "But that doesn't change who you are. You made a mistake. A terrible one. But you're still you. And you're still here."

Her voice didn't waver. She didn't try to console me with clichés. She offered something far more powerful: the reminder that I was still a person worth caring about.

In that moment, I felt the unmistakable presence of grace, not the kind that comes from chaplains or sermons, but the kind God sends through people. Danielle didn't try to absolve me. She didn't offer easy words. She simply sat with me, and somehow, that was enough. In her steady presence, I saw a reflection of the God I had stopped praying to. I was reminded that even in the depths of guilt, I wasn't beyond reach.

The charge of Involuntary Manslaughter, under Article 119 of the Uniform Code of Military Justice (UCMJ), is applied when an individual's unlawful act or omission causes the death of another person, without specific intent to kill or inflict serious bodily harm. It typically involves what the law defines as culpable negligence, which is a degree of carelessness so extreme that it shows a reckless disregard for the foreseeable consequences of one's actions. In some cases, this charge also applies when a person engages in a dangerous act directly affecting another human being, and that act results in death.

When I learned that the Army had formally charged me under Article 119, it hit me with a force I wasn't ready for. It didn't matter that I had no intention of harming anyone, my failure to act responsibly had caused the loss of a life. That truth, undeniable and devastating, would now follow me into a courtroom. I also faced two additional charges for dereliction of duty, compounding the weight of what was already too heavy to carry.

The possible consequences were steep and final: a maximum of ten years confinement, a dishonorable discharge from the United States Army, a reduction to the lowest enlisted rank, and complete forfeiture of all pay and allowances. I remember reading those punishments and feeling the ground shift beneath me. While I had already been living in a state of grief and guilt, seeing the Army's official recognition of my culpability made the shame cut even deeper. These weren't just legal charges, they were a reflection of the irreversible pain my actions had caused.

Between the charge of involuntary manslaughter and the two counts of dereliction of duty, I was facing a maximum sentence of 11 years of confinement at Fort Leavenworth, Kansas. That reality hit me like a freight train. It was crushing, sobering, and yet, deep down, I felt like I deserved worse. Still, CPT Ku, ever the professional and advocate, urged me not to lose heart. She informed me that she was submitting a formal request to the government to consider downgrading the involuntary manslaughter charge to negligent homicide. While I struggled to see the difference emotionally, legally, it was a significant shift. To me, the charge, the punishment, the shame, they all felt the same. I had taken a life through negligence, and no legal terminology could soften that truth. But CPT Ku was doing her job faithfully and diligently. She was looking out for my best interests, even when I couldn't see the value in doing so myself. She drafted the counteroffer, hoping that the judge might accept a lesser charge of negligent homicide. Whether or not it would be accepted remained to be seen, but for her, it was a step toward justice that acknowledged both the weight of the tragedy and my intent, or lack thereof. Negligent homicide—defined under Article 134 of the Uniform Code of Military Justice (UCMJ)—is one of those terms that sounds clinical, almost cold in its precision. But behind that sterile language lies a truth that is anything but detached: someone died. And someone—me—was responsible.

Not intentionally. Not with malice. But through carelessness. Through a failure to be what I had sworn to be: careful, responsible, alert.

Negligent homicide means a death occurred because someone didn't take enough care. There doesn't have to be evil in the heart, only a moment of inattention, a moment when the standards were not upheld. God knows my heart. He knows I never meant to hurt SPC Joseph Tackett. But that doesn't undo the fact that I did. That moment, that mistake, took everything. His life. His future. And it cast a long, dark shadow over mine.

This charge fell under Article 134, one of the broadest and most sobering articles of the UCMJ. It covers actions that are prejudicial to good order and discipline, or that bring discredit upon the armed forces. I now understand, fully and painfully, how my actions did both. And no matter how many times I replay the events, hoping to change something, just for one second, I can't. That is the helplessness that eats away at you when you've taken something you can never return.

The court didn't need to prove that I meant harm. That was never in question. What they needed to prove and what I already knew was that my negligence had cost a man his life. That I failed to exercise the basic caution and responsibility that every soldier, and especially every officer, is trained to uphold.

The penalties for such a conviction are severe: confinement, a bad-conduct discharge, and forfeiture of pay and allowances. But none of those potential punishments could compare to the sentence I had already begun serving, the one of living with what I had done.

When CPT Ku told me that my General Court-Martial would be held on Wednesday, 31 August 2005, at Camp Victory, I didn't flinch. That date was already branded into my future the moment the trigger was pulled. I wasn't afraid of the courtroom. I was afraid of facing myself.

7

MORNING OF JUDGMENT

GENERAL COURTS MARTIAL, 31 AUGUST 2005

Psalm 24:3:

(New Living Translation). Even when I walk through the darkest valley, I will not be afraid, for you are close beside me. Your rod and your staff protect and comfort me.

On the morning of 31 August 2005, I woke up trembling. I hadn't slept much the night before, truthfully, I hadn't truly rested in weeks. That night had been especially rough. It felt like the last night of freedom I would ever have, the final page before the sentencing chapter of my life began. I believed then that I might be sent away for nearly four years, possibly longer. Honestly, a part of me thought I should be sent to Fort Leavenworth for life.

"A life for a life," I kept thinking. "Isn't that justice, Lord? Isn't that what the Bible says?" Maybe that's what I deserve, for all the pain I've caused the Tackett family.

I sat up in bed slowly, feet planted on the cold floor, motionless for several minutes. The weight of it all, my guilt, my shame, the uncertainty of the day, crushed down on me. There was only one thing left to do.

I dropped to my knees and prayed.

"Dear God," I began, my voice trembling, hollowed out by sorrow. "Please guide me through this. I don't know what's coming, and I don't know if I'm worthy of surviving it. But I know I was wrong. I was negligent. And now I must pay for what I've done. Please, Lord, forgive me. If I must go through this, then give me the strength to endure it."

When the prayer ended, I stood and moved with heavy limbs toward the shower. I tried to focus on small, familiar motions, gathering my toiletries, turning on the water, stepping under the flow, but even in the ordinary, there was no comfort.

I cried.

Not quietly, not restrained, but openly and deeply. I cried for Specialist Tackett, for his family, for the wife he left behind. I cried for my own family, my mother, my father, Aunt Shirley, and my wife, and the pain they too were forced to carry because of my mistake. I wondered what they were thinking, what they must have felt knowing what was about to happen to me.

A part of me didn't want to leave the shower. In there, at least, everything was predictable. There were no judges, no gavel, no prison walls, just the sound of water, a temporary escape from the chaos I had caused.

Then, the darkest thoughts returned, thoughts that

had crept into my mind before but now shouted.

"Maybe I could grab a weapon, any soldier's rifle. Chamber a round. Pull the trigger. End it."

"Maybe if I placed the muzzle right where the bullet struck Tackett, that would be justice. Maybe that would make it right."

But again, I couldn't do it.

Somewhere, deep down, buried beneath the shame and self-loathing, I found a flicker of resolve. I couldn't take the easy way out. Not today. Not like this. If God had spared me this long, there had to be a reason. There had to be something more I was meant to face, endure, or give back.

So, I stepped out of the shower. I dried off. I dressed.

And I prepared to receive the punishment I believed I truly deserved.

I met up with CPT Stevenson, the Battalion Assistant Operations Officer, at the HHB 4th Brigade, 3rd Infantry Division Headquarters around 0830. He was there to escort me to my General Court-Martial at Camp Victory, Iraq.

"Lt. Davis, how are you doing?" he asked, his voice lined with genuine concern.

"I'm fine, Sir, considering the circumstances," I replied, trying to maintain composure.

Out of the corner of my eye, I saw LT Watson in the

distance. Something about her presence gave me a sliver of hope, maybe everything would turn out alright. But before I could dwell on the thought, CPT Stevenson turned to me again.

"Let's go inside, Lieutenant," he said.

We entered the HHB Company Headquarters, where he immediately dismissed all junior soldiers and non-commissioned officers, except SFC Rodriguez. The room suddenly felt colder, heavier.

"Lt. Davis, as per Army regulations, I have to do this," Stevenson said, clearly uncomfortable with what was about to happen.

He pulled out a set of handcuffs and began to read me my rights.

"Lieutenant Davis, you have been charged with a crime punishable under the Uniform Code of Military Justice. You are now in the custody of the United States military, en route to General Court-Martial."

His voice was firm, but his eyes reflected regret.

"Put out your hands, Lieutenant," he instructed.

I obeyed without hesitation.

As he fastened the cuffs around my wrists, he asked, "Are they too tight, Lieutenant?"

"No, Sir, they're fine," I replied, my voice cracking

slightly as I fought back tears.

"Lieutenant," he continued gently, "I'm going to place this handkerchief over your hands. I have no intention of embarrassing you, Willie."

"Thank you, Sir. I appreciate that very much," I said softly.

Once the white cloth was draped over the cuffs, SFC Rodriguez approached. He lightly took my right arm and said, with quiet assurance, "Let's go, Sir. It's going to be alright."

CPT Stevenson took my left arm, and together they guided me outside toward three waiting Humvees. As we rounded the second vehicle, I caught a glimpse of LT Watson again. She gave me a soft smile, but when her eyes fell on the cloth covering my wrists, her smile faded.

I lowered my head, overcome with shame. Inside my sunglasses and beneath my Kevlar, I was crying. On the outside, I looked calm. Inside, I was crumbling.

I stepped into the front passenger seat of the Humvee, swung my other leg in, and sat silently as SFC Rodriguez secured the door.

"Vanguard X-ray... this is Knight Six. Departing FOB Prosperity for Camp Victory," I heard CPT Key say over the comms.

"Here we go," I thought to myself, feeling the bottom fall out of my stomach.

"God help me. Please, just help me," I prayed inwardly.

My dream, my career, and my life as I knew it, everything I had worked hard for was now unraveling.

"In just a couple of hours," I thought, "my life will be over."

We arrived at Camp Victory about fifteen minutes later. The route was familiar: straight out Gate 2, a short ride across Route Aeros, a left onto Route Irish, and then on to FOB Victory.

The convoy pulled up in front of the Trial Defense Services building so I could meet with CPT Ku, my assigned defense counsel. CPT Stevenson exited the Humvee first, then opened the door for me and helped me out. He guided me into the Judge Advocate General's Office.

"We'll see you at the court-martial, Lieutenant. Good luck," he said, removing the handcuffs.

As he left, I was greeted by CPT Ku's assistant, an older Sergeant First Class of Latino descent.

"How are you doing, Sir?" he asked, motioning for me to follow him.

"Fine, I guess," I replied.

I walked down a short hallway toward CPT Ku's office. As I approached the doorway, she stepped out to greet me.

"Lieutenant Davis," she said, "how are you holding up?"

"They cuffed me, Ma'am," I said, trying not to sound bitter. "In a combat zone. Unarmed. Why?"

Her expression shifted instantly to one of visible frustration. She strapped on her flak vest, slung her weapon, and shook her head.

"Are you ready for this, Lieutenant Davis?"

"I think I am, Ma'am," I replied, numb.

"Just be yourself," she said. "You'll be fine."

Colonel Patrick Reinert entered the courtroom, and immediately, every person in the room rose to the position of attention. A silence fell, sharp, tense, and absolute.

The weight of the moment pressed down on my shoulders like a rucksack filled with stone. My heart pounded in my chest as if trying to escape the consequences I was about to face. This was it.

"This Article 39(a) session is now called to order," Colonel Reinert stated, his voice calm but authoritative, slicing through the heavy air like a saber.

I stood there in uniform, still and silent, every fiber

of my being trembling under the surface. My career, my future, my very identity as a soldier, all of it hung in the balance in that sterile, dimly lit courtroom.

At the government table, Captain Darren Pohlmann rose slowly and deliberately, flanked by Captain Robert Guillen. Pohlmann adjusted his uniform with precision, then addressed the judge with a rehearsed yet weighty tone.

"This court-martial is convened under Court-Martial Convening Order Number 6, Headquarters, 3rd Infantry Division, dated 18 April 2005, and Order Number 12, same headquarters, dated 22 August 2005. Copies of both have been furnished to the military judge, counsel, and the accused. They will be entered into the record at this time."

He glanced briefly in my direction, not with malice, but with the impersonal resolve of someone fulfilling their duty. My throat tightened.

"The charges have been properly referred to this court for trial and were served on the accused on 3 August 2005. The government is prepared to proceed with arraignment in the matter of United States vs. First Lieutenant Willie Davis."

There it was. My name is spoken not as a badge of honor but as a defendant. In that moment, I was no longer a leader of soldiers. I was a man on trial for an irreversible mistake that had taken a life.

And as those words echoed in the courtroom, all I could think about was Sergeant Tackett, his face, his family, the promise of a future he would never get to see. And mine, now fractured, trembling beneath the weight of guilt and uniform.

For the next ten minutes or so, the courtroom moved with slow, deliberate purpose as the Trial Counsel and my defense counsel, CPT Fansu Ku, walked through the preliminary proceedings. Their voices, steady, practiced, procedural, filled the space, but all I could hear was the thundering of my own heartbeat. Every sentence, every legal formality, felt like another brick in the wall closing in around me. This wasn't just protocol; it was the formal beginning of my reckoning.

Colonel Patrick Reinert, the military judge, sat with the calm command of someone who had seen this scene unfold too many times before. When he turned his attention to CPT Darren Pohlmann, the lead Trial Counsel, my stomach twisted with dread.

"Sir," CPT Pohlmann began, standing with military precision, "the general nature of the charges in this case is one count of involuntary manslaughter, in violation of Article 119; and two counts of dereliction of duty, in violation of Article 92 of the Uniform Code of Military Justice. The charges were preferred by Captain David Key and forwarded with a recommendation as to disposition by Colonel Vincent Quarles and Colonel Edward C. Cardon.

Hearing the charges laid out in that cold, clinical language, "involuntary manslaughter," "dereliction of duty" was like being gutted in slow motion. These weren't just legal terms to me. They were reminders of the worst mistake of my life and the life that had been lost because of it.

When the preliminaries were complete, Colonel Reinert turned his gaze directly at me. His voice, though measured, held an unmistakable gravity.

"Lieutenant Davis," he said, "I'm going to explain the elements of the offenses to which you have pled guilty. By elements, I mean those facts which the prosecution would have to prove beyond a reasonable doubt before you could be found guilty, had you pleaded not guilty. As I state each element, I want you to ask yourself two questions: First, is the element true? And second, do you wish to admit that it is true?"

He paused, letting the weight of his words sink in.

"After I list the elements for you," he continued, "be prepared to talk to me about the facts regarding the offenses."

I sat there, motionless, the uniform I once wore with pride now heavy with shame. This wasn't just a legal procedure; it was the public acknowledgment of my deepest failure. And soon, I would speak the truth of it all before God, the court, and myself.

After a brief silence, he continued, "Do you have a copy of the charge sheet in front of you?" he replied. "Yes, Sir," I replied. "Take a look at Specification 1 of Charge 1, which, as your counsel has pled for you, alleges a violation of dereliction of duty through negligence…. Do you understand that?" "Yes, Sir, I do," I replied.

"This is a violation of Article 92 of the Uniform Code of Military Justice. The elements of this offense are: that you had a certain prescribed duty or duties, and in this case it was to clear your weapon upon entering a forward operating base; second, that you knew or reasonably should have known of the assigned duty; and the third element is that on 23 June 2005, at, FOB Honor, Baghdad, Iraq, you were derelict in the performance of those duties by failing to clear your assigned weapon upon entering the FOB.

In considering the elements, you should also consider those elements and definitions. Consider the following definitions, as well. A duty may be imposed by a regulation, lawful order, or custom of the service. A person is derelict in the performance of a duty when he willfully or negligently fails to perform it, or when he performs it in a culpably inefficient manner.

Dereliction is defined as a failure in duty, a shortcoming, or a delinquency. The term "negligently" means an act or failure to act by a person under a duty to use due care, which demonstrates a lack of care for the property of others or the safety of others, which a reasonably prudent person would have used under the same or similar circumstances.

Why don't you tell me what you did that caused you to enter a plea of guilty to this specification?

"Sir," I stated." In this specification, I did not properly clear my weapon. In so doing, when I entered upon FOB Honor on that day, after my mission, my driver had taken all three of our weapons: my weapon, his weapon, and my gunner's weapon, his M-16, and he would clear them as I would stand between the truck and the actual clearing barrel and verify. While he was clearing the weapons, I was looking at my gunner, making sure his M240B, his crew served weapon on the truck, making sure that he was clearing that.

"In so doing, when I was not paying attention to my own weapon being cleared and actually looking at it. I was looking at him at the same time, looking up at my gunner up in the turret. That is why I feel that I am guilty for this charge."

"So, let's kind of break that down a little bit. You've told me that you had a duty to clear your weapon. How was that duty established?"

"By actually clearing our weapons, Sir, we are supposed to take each of our weapons, put the actual muzzle inside the clearing barrel, clear our weapon with a verifier, making sure that we pull the charging handle back to the rear, a round should then pop out. If it does not pop out, we should look in the chamber and visually inspect it. I am actually being a clearer or the verifier, so you had two soldiers that is verifying to make sure there is no round in the chamber."

"So, you're supposed to send two people up to the clearing barrel at a time?

"Yes, Sir," I replied.

"And in this case, and that's established by the custom of the service, or is there a specific order that's been promulgated to say that you need to have a weapon cleared when you come on the forward operating base?" he continued.

"Yes, Sir."

"So, both by an order, as well as by custom of the

service?" he continued.

"Yes, Sir."

"And you said that you know of this duty?"

"Yes, Sir."

"And you said when you come back from a mission. Was that on June 23rd?

"Yes, Sir."

"And you were entering FOB Honor, here in Baghdad, Iraq? And in this case, you handed your personal weapon, which is what kind of weapon, Lieutenant?"

"A M-16A4, Sir," I replied.

"So, you handed your M-16A4 to your driver, and he was going to clear the weapon for you?

"Yes, Sir."

"Now, routinely, your soldiers are supposed to clear their own weapons. Is that correct?

"Yes, Sir."

"But in this case, you handed it to someone else to have them perform your duty of clearing the weapon?

"Yes, Sir."

"And you did not accompany the soldier to the clearing barrel to observe the clearing of any of the weapons, including your own. Is that correct?"

"Negative, Sir. I was actually in the vicinity, Sir. My

gunner was taking out his bolt in his M240B and taking out the rounds to make sure that it was clear, so when he was actually clearing all of our weapons, I was making sure that my gunner's weapon was clear. So, I was between him and the truck. He was to my right, the truck was to my left, and I was in between. I was at the clearing barrel. I just did not take the scrutiny that I needed to ensure my weapon was cleared."

"So how far were you from the clearing barrel when your driver was clearing your weapon?"

"Probably about 3 feet, Sir."

"And you were watching some... You weren't watching the soldier clearing the weapons into the clearing barrel. You were paying attention to another soldier?"

"I was paying attention to another soldier," I replied.

"Do you agree that your act in dividing your attention and not paying attention to the clearing of the weapon at the clearing barrel was negligent?"

"Yes, Sir," I said.

"Do you agree that a reasonably prudent person would have either cleared their own weapon at the barrel or watched the clearing of the weapon into the barrel and then watched someone else clear the weapon, instead of trying to multitask?"

"Yes, Sir."

"Do you also agree that a reasonably prudent person would have known to either clear their weapon or clear the individual weapons and crew-served weapons sequentially, instead of simultaneously?"

"Yes, Sir."

"And do you agree that you did not act in a reasonably prudent manner in clearing the weapons in this manner?

"Yes, Sir."

"Did anyone authorize you not to perform your duties in clearing your assigned weapon upon entering the forward operating base?

"No, Sir."

"Did anyone or anything force you to fail to perform your duties?"

"No, Sir."

"Do you believe you had any legal justification or excuse for what you did?"

"No, Sir."

"Could you have avoided your actions in failing to perform your duties if you had wanted to? Could you have avoided committing this offense?"

"Yes, Sir.

"What could you have done to prevent committing this offense?"

"Actually, took my weapon, had my driver verify, and cleared my weapon myself."

"Do you believe and do you admit that on 23 June 2005, at or near FOB Honor, Baghdad, Iraq, you knew or should have known of your duties and were derelict in the performance of your duties, in that you negligently failed to clear your assigned weapon upon entering a forward operating base, as it was your duty to do so?"

"Yes, Sir, I do."

After this, I sat back in my chair and turned to look behind me. That's when I saw their faces, friends, fellow officers, and soldiers I had once led. I recognized so many of them: 1LT Danille Smith, 1LT Watson, CPT Martin, LTC Pinnell, most of the soldiers from my own platoon, and even a few from SPC Tackett's. They had come to support me during what was easily the most painful and humiliating moment of my life.

But it wasn't just their presence that struck me, it was the expressions on their faces. Disappointment. Sorrow. Shock. Some tried to hide it. Others couldn't. Their eyes told stories words never could. I felt a deep, cold sensation crawl through my stomach. The shame, the regret, the overwhelming desire to disappear, it all hit me like a crashing wave.

Then, as I scanned the courtroom, my eyes locked with those of 1LT Watson. She didn't look at me with judgment or disdain. Her gaze was steady, reassuring, like a hand on my shoulder. She was telling me, without a word, that everything was going to be okay. Earlier that morning, she had prayed with me, offering comfort in a moment of desperate need. Now, she was here, sitting in support of me, standing in the gap where my wife couldn't be. That gesture, that kindness, meant the world to me. She didn't have to be there, but she was. And for that, I will be forever grateful.

My head snapped forward again as Colonel Reinert resumed speaking.

"Let's turn now to the second specification of Charge I," he said, his voice calm but firm. "This is a slightly different violation of Article 92. It concerns willful dereliction of duty."

He paused, letting the weight of his words settle into the courtroom like dust.

"This charge has three elements of proof," he continued. "First, that you had a specific, prescribed duty. In this case, it was your obligation to exercise proper muzzle awareness with your assigned weapon. Second, that you actually knew of this duty. And third, that on or about 23 June 2005, at or near FOB Honor in Baghdad, Iraq, you were derelict in performing that duty by failing to exercise muzzle awareness."

He looked directly at me.

"Just like before," the judge continued, "this duty may be established by regulation, lawful order, or custom of the service. Lieutenant Davis, do you agree that you had a duty to exercise muzzle awareness?"

"Yes, Sir," I responded, my voice steady but heavy with regret. I knew without question that I had a duty to maintain constant awareness of my weapon's direction. Muzzle awareness means always knowing where the barrel, the business end, is pointed, and ensuring it is never aimed in a way that could endanger others. On that day, I failed in that duty. Not just a little, but grossly. Shamefully.

"And how was that duty imposed upon you?"

"At all times, a soldier is expected to make absolutely sure that their weapon's muzzle is pointed toward the ground," I replied, "and under no circumstances should it ever be directed at another soldier or human being, unless engaged in an actual combat situation."

"And was that duty imposed by regulation, lawful order, or custom of the service?"

"All of them, Sir," I said solemnly. "Custom of the service. Regulation. Lawful orders. Every part of my training and leadership demanded it. And I failed to uphold that."

"A person is derelict in the performance of duty when he willfully fails to perform those duties. Dereliction is defined as a failure in duty, a shortcoming, or a delinquency. Did you fail to perform your duties with regard to muzzle awareness?

"Yes, Sir."

"What exactly did you do or fail to do that made you derelict in the performance of your duties?"

"Yes, Sir. As I was coming through the hall en route to the Lieutenant's hooch, passing through the soldiers as I had many times before, the way that I had my weapon on my tactical sling was mostly parallel to the ground. I have tried to fix it before, but my tactical sling will...it will always be in a position parallel to the ground, and I had my hand on the actual butt stock. As I was proceeding through the hallway, the soldiers would sit on little camp stools, about maybe a foot off the ground, and totaled to be about three feet off the ground, and when I was walking by, I would inadvertently flag my soldiers.

"What do you mean by 'flag,' your soldiers?"

"My muzzle would be in their exact...it would be a danger to them in their vicinity. It would be pointing toward their direction, even though it was kind of pointed that way if I was walking, slanting my hand at a slight angle with fingers pointing downward. If they were sitting right there, it was a danger to them and their safety."

"And the tactical sling, that is an adjustable feature on a rifle. Is that correct?"

"Yes, Sir."

"And many times you'll see people walking around and they've got their tactical slings on the weapon is put over their back, and the muzzle's pointed directly at the ground. Is that correct?

"Yes. My muzzle was always supposed to be pointed at the ground. In this case, it was always perpendicular to my...it was kind of bisecting my and always parallel to the ground."

"So, the way you had yours adjusted, your muzzle was always pointing across the horizon, directly at individuals that you might be walking past. Is that correct?

"Roger, Sir, especially if they were sitting."

"And this event, when you walked by your soldiers going through the hallway, when did that occur?"

That would usually occur every night after receiving

my mission brief for the next day. I would pass them…they would be about, maybe three feet from the lieutenant's hooch, and I would always pass them every night. They would be sitting on both sides. It would be like five or six soldiers on one side and five or six on the other."

"Now, on this particular occasion that we are talking about, which is charged, did that occur on the 23rd of June 2005?"

"Yes, Sir."

"And when you were walking by your soldiers, did you take any kind of step to grab your weapon?"

"Yes, Sir, I did take a step to grab my weapon."

"And did you try to point the muzzle down or did you leave it just resting there, so it was still pointed at your soldiers?"

"I was actually trying to point the muzzle down, but my sling, I would have to kind of cock my sling, and I did not cock my sling toward the ground. Because of the way it was adjusted onto my weapon."

"And why did you, as you call it, 'flag' your soldiers or point your muzzle at your soldiers when you walked by?"

"As far as the flag," I said nervously. My muzzle should have been down, and it was not. I have tried before to kind of mess with my tactical sling and make sure to do it, but I guess I was going to get to my room and do it, so I did not…I was not thinking at the time when I didn't, at right then and there, try to make an effort to adjust my tactical sling and make sure that my muzzle was pointed down to the ground.

"The term 'willfully' means intentionally. It refers to doing an act knowingly and purposely; specifically, intending the natural and probable consequences of the act. Now, you have entered a plea to this specification that you have violated this custom of service and were derelict in a willful manner. Now, what you're telling me thus far may not be willful, so I want you to tell me, how was your conduct willful?

Sir, I knew that my weapon should not have been pointed in their direction, and as I was walking through them, I was approached by one of the specialists who said, "Sir, you need to kind of watch your…you need to kind of watch where your muzzle's pointed.' I did not make the adjustment. I believe that I did not make the adjustment, because I was thinking in my mind that my weapon was clear of any round, and I was not going to shoot or do anything in that sense. I knew of the duty, Sir, and it did go through my mind, and I just chose to keep it there."

"So even though you knew of the custom of muzzle

awareness and an E-4 had tried to correct your conduct and reminded you of that muzzle awareness, that you still knowingly and purposely allowed your weapon to point at other soldiers?"

"Yes, Sir," I replied.

"And you did that because you didn't want to try and adjust your sling there, and you were going to do it later?"

"I was going to do it right, I was thinking to myself, my room was right there. I was going to fix it then because it was irritating me."

His honor looked at me and said, "Do you understand all the elements and definitions that we've already talked about?"

"Yes, Sir."

"Do you understand that your plea of guilty admits that these elements accurately describe what you did?"

"Yes, Sir."

"Do you believe and admit that the elements and definitions taken together correctly describe what you did?"

"Yes, Sir."

"Do you admit that you had a prescribed duty, that is, to maintain muzzle awareness of your weapon on the 23rd of June 2005, and that you actually knew of that assigned duty?

"Yes, Sir."

"Do you admit that on the 23rd of June 2005, at FOB Honor in Baghdad, Iraq, you were derelict in the performance of those duties by failing to maintain your muzzle awareness?"

"Could you have avoided failing to perform your duties if you had wanted to?"

"Yes, Sir."

"You told me a little bit about your tactical sling and the difficulty you were having adjusting it. Do you feel that it was impossible for you to adjust that sling properly to make sure that you would have that muzzle pointed in an appropriate manner?"

"No, Sir."

"And had you, in fact, carried it on other occasions and had it pointed in the appropriate manner?"

"Yes, Sir."

"It was just on this day, as well, as you have told me, apparently, on some other days, that you did not have it adjusted appropriately. Is that correct?"

"Yes, Sir."

"Do you believe and do you admit that on 23 June 2005, at FOB Honor in Baghdad, Iraq, you knew of your prescribed duties and that you were derelict in the performance of those duties, in that you willfully failed to exercise muzzle awareness of your assigned weapon, as it was your duty to do so by custom of the service?"

"Yes, Sir."

"Let's turn, now, to Charge II. This charge is a violation of Article 119, but you have entered a plea of guilty to the named lesser-included offense of negligent homicide, which is in violation of Article 134 of the Uniform Code of Military Justice. This offense has five elements of proof that would have to be proven beyond a reasonable doubt, by the government, by legal and competent evidence, if you had entered a plea of not guilty. These elements are: one that Specialist Joseph M. Tackett is dead; second, that his death resulted from your act, that is, shooting him in the head with an M-4 rifle on the 23rd of June 2005, at or near FOB Honor in Baghdad, Iraq; third, that the killing by you was unlawful; four, that your act, which caused the death, amounted to simple negligence; and five, that under the circumstances, your conduct was to the prejudice of good order and discipline of the armed forces, or was of a nature to bring discredit upon the armed forces…now I, note, Counsel, when I went through the elements, the charge sheet says it's Specialist Tackett, but the stipulation of fact says Sergeant Tackett.

"Sir, he was posthumously promoted, Sir."

"And it's actually he was shot with an M-16 rifle, not an M-4."

"So, is that going to be part of the exceptions and substitutions?"

"Lieutenant, what we're talking about is the particular weapon described in the specification, which indicates it's an M-4, when in fact, it is an M-16 rifle. Is that correct?"

"Yes, Sir."

"And those are going to be...When I talked to your attorney about exceptions, when you entered a plea, you did it by an ...entered a plea to a named known lesser...included offense, which basically changes a couple of the elements, but does not change any of the particular facts. So, what we're going to do in making my findings, since we've not determined that it is, in fact, an M-16 rifle and not an M-4, I will make exceptions and substitutions to except out the word "M-4 and substitute, therefore, the word "M-16.""

"Yes, Sir."

"But in essence, if you're dealing with an M-4 or an M-16, what are the differences between those weapons?"

"The muzzle on my M-16A4 is longer. The M-4 is shorter. My M-16 was much longer than the dimensions of the weapon's muzzle of an M-4."

"And the M-4 is more of the carbine version of the M-16."

"Roger. Yes, Sir."

"But it still shoots the same caliber round, a 5.56. Is that right?"

"Yes, Sir."

"Now, in considering the elements, you should use and understand the following definitions. Conduct prejudicial to good order and discipline is conduct that causes a reasonable, direct, and obvious injury to good order and discipline. Service discrediting conduct is conduct that tends to harm the reputation of the service or lower it in public esteem.

The killing of a human being is unlawful when it is done without legal justification or excuse. Simple negligence is the absence of due care; that is, it is an act or a failure to act by a person who is under a duty to use due care, which demonstrates a lack of care for the safety of others, which a reasonably careful person would have used under the same or similar circumstances. The act or failure to act must not only amount to simple negligence but must also be the proximate cause of death. This means that the death of Sergeant Joseph Tackett, who in the specification is noted as a specialist, must have been the natural and probable result of your negligent action or failure to act."

"Do you understand that?"

"Yes, Sir, I understand."

"Do you understand the elements and definitions as I have read them to you?"

"No, Sir, I have no questions."

"Do you understand that your plea of guilty admits that these elements accurately describe what you did?"

"Yes, Sir."

"Do you believe and admit that the elements and definitions taken together correctly describe what you did?"

"Yes, Sir."

"And the name of the deceased is Specialist Joseph M. Tackett, posthumously promoted to Sergeant Tackett?"

"Yes, Sir."

"Why don't you tell me about what occurred on the 23rd of June 2005 with regard to Specialist, not Sergeant Tackett?"

"Yes, Sir. I was going back to my room, my hooch, coming from a meeting with my battery commander, trying to prepare… I was actually going to prepare for the next day's mission. As I was approached…as I was going through the platoon area of operations, the 7th platoon AO, I went through, and they would…on a normal basis, they would stop me and say hi and joke around with me a little bit. On this particular day, they did all of that. They stopped me, said hi, and I was approached by Sergeant…I'm sorry, Specialist Guile, who was sitting down at the time. He said. "Sir, please take more…pay attention to your muzzle awareness of the weapon, because you know how lieutenants like to have negligent discharges." And when I was actually going through, I was pointing my weapon or flagging my soldiers on my left side. Once Specialist Guile told me or said that statement, I immediately turned around and I said, "No, I will not get a negligent discharge. That was your old lieutenant, and that was another lieutenant in my battery who got a negligent discharge. It won't be me." And there was laughter, and soldiers were talking among themselves and doing what soldiers do. They were preparing for their platoon leader and their platoon sergeant to come back from their meeting.

They would affectionately call me one of their nicknames. All of us had them. They called me "Aflac." I don't know where they got it from, but…

"You mean AFLAC, like the insurance company and the duck commercial?"

"Yes, Sir."

"Okay."

"Lieutenant...one of the soldiers made a comment and said, "ducks don't know how to handle weapons properly.'" I don't know where the comment came from. Not only did I have a bad habit of keeping my weapon parallel to the ground and not ensuring that my muzzle was not pointed towards the ground. When I was going through the actual platoon area and my hand was on the butt stock, I was flicking my safety selector switch from safe to semi, back and forth. It was something I did on a regular basis. I don't know...I hate to say it, but it probably had to do something about what I was thinking about at the time, whether it had to do something with the mission. It was such a bad habit that it became, I hate to say it, but it became second nature.

My weapon was pointed in Sergeant Tackett's direction. When I was trying to move the selector switch from safe to semi after realizing the error that I had made, my trigger finger was on the trigger. When I was moving it...trying to move it back to safe, my trigger finger jolted the trigger, and my weapon fired. I looked down at my weapon. I was wondering why it fired. I was shocked. I thought that it was clear of any round.

One of the soldiers had run and gotten up after all the soldiers had.... most of the soldiers had run and cleared out and went their various different ways; they said, "What, Sir? How did that happen?" That comment was actually, "What the fuck, Sir?" I looked at him with…this PFC Ketteler, I looked at him and then I looked towards Tackett. Tackett was sitting in direct…he was sitting in direct…he was sitting." It was becoming difficult for me to say because it was hard to tell the story and realize how reckless I was.

"Take your time." The judge instructed me as he saw me becoming emotional.

"He was in front of," I continued. "He was…my weapon was… my muzzle was pointed in his direction, like I said, because my weapon was parallel to the ground. He was sitting on a camp stool; one of those little field artillery camp stools about a foot high. I looked at him. He had taken his right forearm and he put it across his chest. He immediately tilted down and kind of towards the side, and I screamed his name; I said, "Tackett!" And then when he hit…when his head hit the ground, I saw that I had shot him in the head and blood was coming out on the floor.

When I looked at that, I didn't know what to do. I ran over to him, straddled him, and placed my hand on his head as I tried to apply pressure to his head. I realized that the bullet not only went through one side, but it went out of the other. And after that, I became completely hysterical." "So, when you talked about your hand being on the butt stock of the weapon, what you're talking about is like the pistol grip portion of the M-16?"

"Yes, Sir."

"Where would it normally be if you were going to fire the weapon?"

"Yes, Sir."

"And in this case, you were flipping the selector switch between safety and semi on the M-16, and you also had your finger inside the trigger well of the weapon?"

"Yes, Sir."

"So, your finger was next to the trigger?"

"Yes, Sir."

"And this took place at FOB Honor, here in Baghdad, Iraq?"

"Yes, Sir."

"Was your handling of the weapon on that day, that you've described to me, having your hand on the pistol grip, flipping the selector switch back and forth with your finger inside the trigger well; was that what a reasonably careful person would have done under the same or similar circumstances?"

"No, Sir."

"What would have a reasonably prudent person done

in handling a weapon in a crowded barracks hallway?"

"Pretty much leaving it alone, keeping the muzzle pointed towards the ground; wouldn't touch the butt stock or anything."

"And if … when you shot the weapon, was that an intentional trigger pull, or was it more that it was your finger that pulled the trigger as you were moving the switch?"

"It was more my finger pulled on the trigger as I moved the switch. There's no way I would intentionally…no way would I intentionally shoot, even if it was in a green status. "Would a reasonably careful and prudent person have their finger anywhere close to the trigger, inside the trigger well, when they're in a crowded hallway of barracks?"

"No, Sir."

"Do you agree that Specialist Tackett's death resulted from your act?"

"Yes, Sir."

"Do you agree that his death was the natural and probable consequence of your act of shooting him in the head?"

"Yes, Sir."

"Do you believe that his…Specialist Tackett's death was caused by anything other than you shooting him in the head?"

"No, Sir."

"What impact does it have for one individual to negligently shoot another individual on good order and discipline or service discrediting? I'll break that down a little bit. What impact does it have on good order and discipline for soldiers to shoot other soldiers?"

"I would hope, seriously, a negative impact."

"Does it make the service function more efficiently or less efficiently if soldiers shoot other soldiers?"

"I would think more efficiently, due to the accident..."

"This may be a lesson learned, but if this were to go unchecked, where it was the law west of the Pecos in the barracks and folks were getting shot all the time, would that make the service function more efficiently or less efficiently?" "And how would it appear to the public? Would the public hold the military in greater esteem or lower esteem if soldiers were randomly shooting other soldiers?"

"Lesser esteem, Sir."

"So do you agree that this conduct by you is both service discrediting and conduct which is prejudicial to the good order and discipline of the armed forces?"

"Yes, Sir."

"You told me about your actions with playing with

the selector switch and pulling the trigger and having your weapon pointed parallel to the ground, as opposed to down at the ground. And kind of putting all those things together caused your weapon to go off and shoot Specialist Tackett in the head. Is that kind of a clear picture of what you've told me so far?"

"Yes, Sir."

"In concert with all of those things together, could you have acted in a manner differently that would have resulted in a different ultimate result, that you would have not shot him in the head?"

"Yes, Sir."

"What kind of choices could you have made that would have prevented Specialist Tackett from being shot in the head?"

"I could have, one, made sure that my weapon was always pointed down to the ground due to my tactical sling and that it wouldn't slip and slide during the day or what not. At that point, my hand would have never been on the butt stock of the weapon at all. Basically, if I had made sure that my weapon was positioned in a way that it should have been positioned, close to my chest, always with the muzzle down, it would have never happened if I had acted on that."

"And you wouldn't have been playing with the selector switch; that would have helped?"

"Yes, Sir."

"If you had actually cleared the weapon, that would have certainly prevented the shooting, right?"

"Yes, Sir."

"And from the stipulation of fact, that there are apparently some other occasions that you may have done some things that were inappropriate in handling the weapons as well; correct?"

"Yes, Sir."

"Do you want to tell me about those?"

"There was one time where I was in the dining facility and for some stupid reason on my part, I did pull my bolt to the rear and kind of, not completely charging it, but just pulled it back to the rear. I forget what incident actually curtailed me to do something like that, but that was not anything that was supposed to be done, even if my weapon was green or cleared at the time."

"And for folks who aren't familiar with the M-16 and for the purposes of the record, when you pull the charging handle of an M-16 back and you pull the bolt to the rear, is that the step that normally called charging the weapon and actually can, if there's a magazine in the well, will actually put a round into the chamber?"

"Yes, Sir."

"And you did that when your weapon was on green status, but you were in the DFAC. Is that correct?"

"Yes, Sir."

"So, these were what you've told me about, are kind of a pattern of some bad habits, with regard to the handling of your assigned weapon. Is that right?"

"Yes, Sir."

"And do you agree that that pattern of mishandling your weapon: flagging your soldiers, pointing the muzzle of your weapon at those soldiers, flipping your selector switch kind of at random, and failing to clear the weapon were all steps that were the proximate cause of the death of Specialist Tackett?"

"Yes, Sir."

"And at any time during that chain of events and those habits that you have developed, you could have voluntarily stopped and changed those habits and created a safer environment?"

"Yes, Sir."

"And that you had the duty to act in a reasonably prudent way and not have all those things take place; correct?"

"Yes, Sir."

"Do you believe you had any legal justification or excuse?"

"No, Sir."

"Did anyone force you or threaten you to get you to engage in this conduct?"

"No, Sir."

"Do you believe, and do you admit that at or near FOB Honor, Baghdad, Iraq, on or about 23 June 2005, you unlawfully killed Specialist Tackett by shooting him in the head?"

I felt my heart literally stop as guilt and shame seemed to cover me as I replied, "Yes, Sir."

"And that your shooting of Specialist Tackett was an act of simple negligence; an act that a reasonably prudent person would not have taken under those circumstances?"

"No, Sir. I mean...."

"That may be a confusing question. Were your acts negligent?"

"Yes, Sir."

"And do you believe, under the circumstances, that your conduct was prejudicial to the good order and discipline of the armed forces?"

"Yes, Sir."

"Was the killing by you unlawful?"

"Yes, Sir."

After this line of questioning, the military judge turned to both the defense and trial counsel and asked if there were any additional questions pertaining to this particular specification. Captain Ku and the trial counsel each respectfully declined. Then I heard the Colonel say, "Let's take about a ten-minute recess so the court reporter can reset his machine... We're in recess."

As the courtroom rose while the Colonel exited, I sat back down, burying my face in my hands. I couldn't help but wish desperately that my life and military career hadn't come to this. Prudent. Reasonably prudent. I was neither of those things on that day. I failed to exercise sound judgment with my weapon. My carelessness, no, my gross imprudence was the sole reason for Sergeant Tackett's death.

I had taken another man's life because I had failed, not just to follow Army regulations, but to correct myself. I was a commissioned officer in the United States Army, for God's sake. A leader. A standard bearer. And yet I failed. Horribly. This should never have happened, I thought, again and again. Not on my watch. Not from my hands.

"Hey, Lt. Davis... you're doing fine," CPT Ku said, her voice pulling me out of the downward spiral in my mind.

I looked up and saw her offering a warm, reassuring smile. It was her way of trying to steady me, to remind me that I wasn't alone in this moment. "Just keep doing what you're doing, and you'll be fine," she said. "Do you have any questions?"

"No, Ma'am... no, I don't," I replied softly.

She gave a small nod and turned to speak with the trial counsel. A moment later, 1LT Smith and 1LT Watson approached me with quiet concern in their eyes. They had come to check on me, and for a moment, just seeing their faces helped me breathe a little easier.

"Hey, Davis, how are you doing, man... You hanging in there?"

1LT Smith asked softly, placing a comforting hand on my shoulder.

"I guess," I replied, my voice barely above a whisper. "I just still can't believe this is happening to me, to all of us. This is a nightmare. How am I supposed to move on with my life when this will always haunt me? The negligence, the bad habits that should've been corrected... I should have corrected them."

"Just have faith in God and pray, Davis," LT Watson added gently. "Through God, everything is possible."

"What the fuck ever," I thought bitterly. I didn't say

it out loud, but the anger stirred in my chest. Why did God allow this to happen in the first place? Why wasn't there some divine intervention to stop me? To stop the weapon? To spare SPC Tackett?

If I was being honest, part of me was angry at God. Angry that He let it happen. Angry that, in my moment of carelessness, no unseen hand intervened. But beneath that anger, I was beginning to see the truth, uncomfortable, raw, and suffocating.

I was trying to place the blame on everything and everyone but myself: my tactical sling, my bad habits, even God. But the blame was mine. All of it. And accepting that was the hardest pill I had ever had to swallow.

In truth, I was scared. Scared of what lay ahead. Scared of the unknown. Scared of the punishment I knew I deserved.

I didn't know where my life was going from here, all I knew was that it would never be the same again.

8
SENTENCING

Psalm 51:3-4:

"For I know my transgressions, and my sin is always before me."

After the short break, the Master of Arms was motioned to prepare the courtroom for the Colonel. "All rise," I heard the Master of Arms command as the Colonel entered through the double doors of the courtroom in a seemingly stoic look and proceeded to the bench to take his seat.

"Court is called to order. The record should reflect that all parties present when we recessed are now, again, present. Does either party feel there's any additional inquiry necessary on any of the charges and specifications?"

"No, Your Honor," I heard both my lawyer and the trial counsel say in response to Your Honor's question.

"Trial Counsel, what do you calculate the maximum punishment to be?"

"Sir, we calculate it at 45 months, Sir. 36 months for Charge II, negligent homicide; and Charge I, Specification I, negligent dereliction, 3 months; and Charge I, Specification II, willful dereliction, is 6 months, Your Honor."

"So, a total of 3 Years, 9 months?"

"Yes, Your Honor."

"Defense Counsel, do you agree?"

"Yes, Your Honor."

"In addition to the confinement, there are also other aspects of the punishment, including dismissal and total forfeitures. Correct."

"Yes, Your Honor."

"Lieutenant Davis, the maximum punishment authorized in this case, based solely on your plea of guilty, is confinement for 3 years, 9 months; a dismissal; and total forfeiture of all pay and allowances, and a fine may be adjudged...

"Do you understand that?"

"Yes, Sir."

"On your plea of guilty alone, this court could sentence you to the maximum punishment, which I have just stated."

"Do you understand that?"

"Yes, Sir."

"Do you have any questions as to the sentence that could be imposed as a result of your guilty plea?"

"No, Sir."

"Lieutenant Willie Davis, Jr., I find that your plea of guilty is made voluntarily and with full knowledge of its meaning and effect. I further find that you have knowingly, intelligently, and consciously waived your rights against self-incrimination, to a trial of the facts by a court-martial, and to be confronted by the witnesses against you. You have further provided a factual basis for the entry of your guilty pleas, both in what you've told me here in court, as well as what is contained in the stipulation of fact.

"Accordingly, your plea of guilty is provident and is accepted. However, I advise you that you may request to withdraw your plea of guilty at any time before the sentence is announced, and if you have a good reason, I will grant it."

"Accused and defense counsel, please rise."

"Lieutenant Willie L. Davis, Jr., in accordance with your pleas of guilty, this court finds you: Of Specification 1 of Charge I: Guilty, except the word 'willfully,' and substituting therefore the word 'negligently,' of Specification 2 of Charge I: Guilty. Of the specification of Charge II: Not Guilty, but Guilty of the lesser-included offense of Negligent Homicide, in violation of Article 134."

"Please, be seated."

"Lieutenant Davis, we now enter the sentencing phase of the trial, where you have the right to present matters about the offenses or yourself which you want me to consider in deciding your sentence. In addition to the testimony of witnesses and the offering of documentary evidence, you may, yourself, testify under oath as to these matters or you may remain silent, in which case I will not draw any adverse inference from your silence." On the other hand, if you desire, you may make an unsworn statement. Because the statement is unsworn, you cannot be cross-examined on it. However, the government may offer evidence to rebut any statement of fact contained in an unsworn statement. An unsworn statement may be made orally, in writing, or both. It may be made by you, by your counsel on your behalf, or by both.

"Do you understand these rights?"

"Yes, Sir, I understand."

After the judge finished explaining what was going to take place during the sentencing phase of my court-martial, I learned something surprising: I had been ill-treated during my transfer to Camp Victory. When CPT Stevenson placed me in handcuffs during the convoy from the Green Zone, it was deemed unnecessary. I wasn't a flight risk. I was in the middle of a war zone. I wasn't armed. Because of this, my defense attorney, CPT Ku, was able to successfully argue for five days of credit toward whatever sentence I might receive. It was a small but symbolic moment, one that acknowledged that even amid my guilt, I had still been treated unfairly.

Still, as the court-martial drew to a close, my nerves frayed. For the first time in my life, I had no idea what was waiting for me, not the next day, not the next week, not the next year. I felt like I was staring into a black hole. All I could do was hope, hope that the Army would dismiss me and send me home to be with my wife. I prayed that the court might see that living with the fact that I had killed SPC Tackett, accident or not, was punishment enough. I would carry that guilt for the rest of my natural life. Every day. Every night. With every breath.

I hoped they'd discharge me, take away my Second Amendment rights, and label me a convicted felon. Maybe that would be enough. Maybe not. But it was the best outcome I dared to hope for.

Yet deep down, I expected the worst. How could I not?

I was a First Lieutenant in the United States Army. I was supposed to be a model of discipline and responsibility, especially when handling a weapon, a tool of war meant to protect, not destroy from within. Officers aren't allowed to be negligent, let alone grossly negligent. What I did violated everything I stood for. Everything I had sworn to uphold.

So yes, I expected the harshest outcome: 45 months

in confinement at Fort Leavenworth, a dismissal from the Army, and the forfeiture of all pay and allowances. And if that was the sentence, then so be it. I believed I deserved it.

Even though I had signed a pretrial agreement limiting the maximum sentence to 36 months, part of me still felt like that wasn't enough to atone for what had happened. Not really. Nothing ever would be.

Now, soldiers from my platoon and others from the battery were being called to the witness stand to testify about what happened that night. As I looked around the courtroom and saw all the familiar faces, my soldiers, fellow officers, and others who had come in support of me, I felt a strange mix of relief and tension tighten in my chest. I didn't know what to expect. Did they hate me for what happened to SPC? Tackett? Did they blame me, even though it was an accident? Or did they feel sympathy?

The truth was, I hadn't spoken to most of the platoon or the battery in nearly two months. I had no idea where they stood, how they felt, or what they thought of me now. I could only sit there, quietly bracing myself, as the people I had once led, brothers and sisters in arms, stepped forward to speak about the worst night of our lives.

"Trial Counsel, you may proceed with direct examination, call your first witness."

"Yes, Sir. The government calls Sergeant First Class Larry Karns, United States Army, as our first witness.

I saw Sergeant Karns stand up and begin to make his way towards the stand. He was the battery Communications NCO and a damn good one, I thought. He was promoted to Sergeant First Class and became the Platoon Sergeant of Seventh Platoon, Alpha Battery. As he approached the bench, he glanced over towards me, nodded his head, and smiled before he was administered the oath by trial counsel. The smile lightened me up and relaxed me a little bit.

"Please state your name and unit for the record."

"Larry Karns, Alpha Battery, 1st Battalion, 76 Field Artillery."

"Please have a seat."

"Sergeant Karns, I note that when you stated your name, you talked a little fast. One thing you need to do for me is slow down a little bit, because every word you say, the court reporter has to say."

"Yes, Sir."

"Sergeant Karns, can you please tell the court how long you have been in your present position?"

"Sir, I have been in my present position since June or July of this year."

"Was it before or after the incident?"

"It was approximately, maybe, 3 weeks before the incident."

"How long have you been associated with the battalion?"

"I have been in the battalion since October of 2004…excuse me, 2005."

"Did you say October?"

"I'm sorry, last year."

"October 2004."

"Yes, Sir."

"How long have you known Sergeant Joseph Tackett?"

"I have known of him before our deployment out here. Once we arrived over here in Iraq, I was a little closer to him and the rest of the platoon; we lived in the same area."

"Did you have an interaction with him on a daily basis or was it like once in a while?"

"Correct. Once we moved into…came into Iraq, it was a day-to-day basis where I'd seen him and interacted…"

"Would you say you knew him pretty well?"

"Yes, Sir."

"How would you describe Sergeant Joseph Tackett in

terms of either his job performance or his personal personality?"

"Sergeant Tackett, job performance wise, for a specialist at the time, was outstanding; very outstanding. He took great initiative, and he took pride in everything that he did. Offline, he was very humorous. He kept the spirits up and kept the guys motivated."

"Can you give us an example of...how do you know that he was humorous and kept spirits up? Do you have example...specific examples."

"There is one individual within the platoon that was a little slow and wasn't really...kind of like an oddball of the platoon, I guess you could say. Tackett had a way of getting him to open up and express himself, and the guys just...he's become one of the regular guys now. He's not kind of left out or shadowed out. He's brought in."

"What was Sergeant Tackett's position at the time?"

"At the time, he was the gun truck leader, squad leader...or team leader."

"For a team leader, usually it's at least an E-5.

"Do you know why the chain of command put him in that position?"

"Roger. The chain of command put him in that

position because he was very tactful, and he knew… he was very knowledgeable, and he handled himself very professionally."

"Did he really fit the role of a team leader?"

"Yes, Sir, he did."

"How was the morale before Sergeant Tackett was shot and killed?"

"The morale of the platoon was particularly good. Everyone was high-spirited. They were laughing and joking, and Sergeant Tackett brought that to them."

"Did Sergeant Tackett's humor help maintain the morale?"

"Yes, Sir."

"Had the unit taken any casualties prior to the shooting of Sergeant Tackett?"

"No, Sir."

"Had they been hit with any IEDs?

"The platoon was hit with a VBIED at one point in time, but other than that, no, Sir."

"Were there any injuries out of that, minor, however?"

"No, Sir."

"So, this was the first injury in 7 platoon, Alpha Battery, 1st of the 1st Battalion, 76th Field Artillery."

"Yes, Sir."

"How would you describe the platoon, close, tight, before the incident?

"Before the incident, they were very tight. A very tight group. They all bonded together and spoke with each other and looked out for one another."

"After Sergeant Tackett was killed, how did the members...what happened to the platoon? Were they just...explain what happened with the platoon after?"

"After the incident, it was a lot of distraught; a lot of not sure of where they are or what they're doing; second guessing themselves and their abilities and what their tasks were." "Did any of them talk to you directly about it or was this just a feeling you got from watching the soldiers?"

"Oh no, they all came to me and spoke with me about it and told me some of the issues they were having; and different ways of how they can deal with some...deal with the incident."

"How long did it take some of the members of the platoon to get over this incident, if they ever did?"

"Roughly, I'd say, probably about a month to a month-and-a-half before everyone was back up on track and had their mind right and ready to get back on the road and focus on what we're doing."

"What happened with the unit as a whole? Was it just thrown…put right back on a mission or what happened?"

"The platoon was off the road, I believe it was 3 or 4 days we were kept off the road after the incident. Then on the 4th day, I went to the battery commander and requested to put the platoon back on the road to get their mind off of what was…of everything."

"What's the unit's mission? What's the platoon's normal job?"

"We escort Embassy personnel and government officials to different locations throughout Baghdad."

"How often did they convoy prior to the incident, on a weekly basis?"

"Once a week…or once a day throughout the week and then they'd rotate into a FOB Security posture and then go back on the road, so they're out every day."

"So, for about 4 days, this entire platoon was taken offline?"

"Yes, Sir."

"And they weren't there to fulfill their duty, to fill…"

"Objection, Your Honor; leading."

"Sustained."

"I have no further questions, Your Honor."

"Defense."

"Yes, Your Honor."

"You've known Lieutenant Davis since October 2004?"

"Yes, Ma'am."

"He has a very good heart, in your opinion?"

"Yes, Ma'am."

"He's a little…he's a young lieutenant?"

"Yes, Ma'am."

"He simply needs a little mentoring?"

"Yes, Ma'am."

"And you've given him mentoring before?"

"Yes, Ma'am."

"Pulled him aside?"

"Yes, Ma'am."

"And he's always been willing to accept any mentoring on your part?"

"Yes, Ma'am."

"Now, you've never personally seen Lieutenant Davis flag any of the soldiers in the platoon; correct?"

"No, Ma'am, I haven't."

"And if members of your platoon had brought that to your attention, would you have pulled Lieutenant Davis aside and corrected him?"

"Yes, Ma'am, I would have."

"Now, you weren't there; you didn't see the shooting of Sergeant Tackett; correct?"

"Correct."

"But you saw Lieutenant Davis afterwards?"

"I'd seen him briefly, yes, Ma'am."

"He was shocked?"

"Yes, Ma'am. He was sitting down with his head in his hands."

Devastated. Would it be fair to say that he was devastated by what happened?"

"It appeared that way, yes, Ma'am."

"Thank You.

"May this witness be permanently excused or temporarily?"

"Permanently excused, Your Honor."

"No objections, Your Honor."

"Sir, the government calls Specialist John Allen, United States Army."

"Please state your name and unit for the record, please."

"Specialist John Allen; Alpha Battery, 1-76 FA Artillery."

"How long have you been associated with the battery?"

"I've been with the battery since they formed the new battalion. Prior to that, we were Charlie 1-41 FA. I've been here for 3 and a half years."

"How long have you known Sergeant Joseph Tackett?"

"3-and-a-half years."

"Were you close with him, or did you just know him as an acquaintance?"

"In the Rear, we were two different Military Occupations Specialties, (MOS). We were FDC; I was field artillery, but over here, we were pretty close."

"Did you serve with Sergeant Tackett during OIF?"

"Yes, we were in the same battery. We were Charlie Battery, 1-41 during OIF I."

"How would you describe Sergeant Joseph Tackett as a soldier?"

"As a soldier, he was a leader. He worked hard. Any mission you gave him; he'd make sure you had good results at the end."

"How was he as a person?"

"As a person, just fun-loving, love life. He liked to joke around a lot. Just a really good guy."

"Do you have any examples to show the court?"

"Can't really think of anything in particular. You

know if you were having a bad day, he was going to make you laugh. He wouldn't let anyone around him have a bad day. If you were…if he was in your presence, you're going to have a good day."

"Do you know anything about his family?"

"Just that he was married. Had been married since he came in the military, and other than that…"

"Do you know where he was from?"

"He was from Kentucky. He and his wife were both from Kentucky."

"How would you describe the platoon?"

"We have a very close-knit platoon. We have a motto, "The Best," and we take pride in that. It's a really close, tight platoon."

"After Sergeant Joseph Tackett was killed, how did the unit take it? How did they take it?"

"It kind of hit hard, just because you kind of expect to lose a life doing your job, you know, out on the road doing patrols, but not in the safety of 'home.' When you're inside the FOB, it's supposed to be safe. It was hard; it was really hard."

"How long did it take you, personally, to come to grips with the loss?"

"Me personally, it was pretty quick, because the

command was looking at me to hold the soldiers together, the younger soldiers. So, I personally put it behind me. I'll deal with it later. I had to be tough and strong for our younger soldiers."

"Did any of the younger soldiers come to you?

"All of them."

"How did they deal with it? How long did it take them to get over it?"

"We had two or three soldiers who had a really hard time with it. They weren't sure what to do, which way to go with it. I tried to lead them. We had one of our NCO's, one of the team leaders, who had a hard time dealing with it; just there to give them support, a shoulder to cry on, whatever they needed."

"On the night of 23 June, where were you sitting relative to Sergeant Tackett?"

"I was sitting directly across from Sergeant Tackett."

"How do you remember Lieutenant Davis's actions on the night?"

"Objection, Your Honor. This is sentencing. We've already determined what the facts were."

"For what purpose is it offered?"

"To show the recklessness, to show how close he was and to see the weapon handling."

146

"I'll allow it for the impact of the …the nature of the events that were observed by this witness go to the impact of the crime on the unit and for all the individuals who were personally present. So, I'll allow it for background."

"Thank you, Your Honor."

"Do you remember Lieutenant Davis's actions on the night of the shooting?"

"He came through the hallway, which he did every night, because his room… he had to go through us to get to his room for his briefing. He walked through that night and kind of flagged a couple of people with his muzzle, and we were, "Hey, Sir, watch your muzzle. Muzzle Awareness." And everyone was just…it was lighthearted, you know; serious, but lighthearted; "Hey, Sir, watch your weapon.""

He walks through. There are approximately 6 people on one side of the hallway and 6 people on the other. He's walking through the middle. He gets to the end and turns, and someone…something was said, someone made a remark, and he turned around and literally pointed…brought his weapon up with one arm and pointed it at three or four different people. Then Sergeant Tackett said something, and he pointed it at Tackett, and you heard the safety…his safety flip to fire; you heard the click, and before anyone said anything, the round went off.

"Do you remember the look on Lieutenant Davis's face?"

"The last look that I'd seen was because I was looking at him when I heard the click, he was smiling because it was in humor. I don't think he was taking it as seriously as everyone else was, and he was smiling. When the round went off, the last thing I'd seen was him smiling, and then Sergeant Tackett fell to the side. His head fell 6 inches from my boot, and the only thing that I could think of was to go get a medic, so I immediately ran out of the hallway to get a medic."

"How has this incident affected you, both professionally and personally?"

"Professionally, I'll definitely learn from it. I will always make sure that not only is my weapon clear, but my fellow soldiers are clear, just more aware of everything. Personally, it was hard. I had a…you know, I'm 35 years old, I have a new baby at home. He wasn't even crawling when I left, and just the thought that it could have been me and I wouldn't have seen my new baby grow up."

"No further question, Your Honor."

"Defense."

Captain Ku stood up and began to question Specialist Allen.

"Specialist Allen, you also saw Lieutenant Davis…you saw his reaction right after Sergeant Tackett was shot; correct? He was extremely shocked."

"When I came back in…After I went and got a medic, came back in, brought the medic to Tackett, they started working on him, trying to stabilize him. As we were leaving, you know, that's when you could tell he was just in total shock."

"Very emotional?"

"Pretty much, yes, Ma'am."

"And I believe, in your own words, he looked like he was trying to climb out of his skin?"

"Yes, Ma'am."

"You don't hold any grudges against Lieutenant Davis, do you?"

"No, Ma'am. It was a tragic accident that shouldn't have happened. If proper clearing procedures had been done, it wouldn't have happened."

"And in your opinion, two lives were lost…"

"Yes."

"On the 23rd of June, Sergeant Tackett, of course. The person responsible?"

"Sergeant Tackett, obviously, is gone. He's no longer here with us. Lieutenant Davis, he's got to live with this the rest of his life."

"Thank you."

"Trial Counsel, anything further?"

"Nothing further, Your Honor."

Private First Class Adam Mattis was called as a witness on my behalf. PFC Mattis, my driver, good soldier, a greater guy. The assistant trial counsel addresses Private Mattis for the record.

"State your name and unit for the record, please?"

"Sir, my name is Private First-Class Adam Michael Mattis. I'm with the 1st Battalion, 76 Field Artillery, Alpha Battery."

Captain Ku began to question Private Mattis. I can tell that he was a little uneasy.

"Private Mattis, do you know Lieutenant Davis?"

"I do, Ma'am."

"And how long have you known him?"

"Ma'am, it'd be about a year."

"And how did you come to know Lieutenant Davis?"

"At first, Ma'am, he was. He was part of my platoon. Then, at JRTC rotation, I was his driver. And from that point on, I was associated with him through different things at the battery."

"Could you speak up a little bit there, Private?"

"I apologize, Sir."

"That microphone won't amplify you, so you need to speak up loud enough so I can hear you. Thank you."

"Roger."

"Can you repeat the last answer?" How did you come to know Lieutenant Davis?"

"Yes, Ma'am. At first, when I arrived in the battery, he was associated with my platoon. At the JRTC rotation, I was his driver. And then, when we got back, I helped him with supplies, and then I was obviously his driver, and he was my platoon leader in the country, Ma'am.

"And how long has he been your platoon leader now?"

"Ma'am, since we got in country, which would have been approximately February 2005."

"Had you been out on missions before with Lieutenant Davis?"

"A good deal, Ma'am."

"And from those times that you've been out on missions with him; does he know the proper clearing procedures?"

"Absolutely, Ma'am."

"And did he follow through with those clearing procedures on the times that you've gone out with him?"

"Ma'am, when I was driving for Lieutenant Davis, we had a routine. When we got back, we'd clear my weapon first. He'd stick his finger in the chamber, then I'd do the same for his, then we'd go and get the gunners."

"How would you describe Lieutenant Davis's leadership style?"

Ma'am, he is unique on his own, in the fact that he respects us in a way that…it's different than other officers. He respects our opinions. He asks us for advice. He asks us how we think things would be done, how we like things to be done. He always takes that into consideration before making his decisions.

"In your opinion, did he take care of his soldiers?"

"Absolutely, Ma'am."

"And would you work with Lieutenant Davis again?"

"Without a doubt, Ma'am."

"Now, did you see him on the night of the 23rd of June 2005?"

"I did, Ma'am."

"And how would you describe his demeanor?"

"Nothing less than hysterical, Ma'am. I've never seen him or anyone else act like that before."

"Could you give us a little bit more detail?"

"Ma'am, I saw him being dragged out of the area by

two NCOs on his knees, sobbing."

"Thank You."

"Trail Counsel."

"You've been in the military 3 plus years, right?"

"That's a negative, Sir."

"How long have you been in the military?"

"Sir, I arrived in my Basic Training on the 5th of February 2004."

"You went to Basic Training?"

"Roger, Sir."

"AIT?"

"Roger."

"You learned about muzzle awareness?"

"Sir, we did, in Basic Training, learn about muzzle awareness."

"It's important to you to be cognizant of where your weapon is pointed at all times?" "Roger, Sir."

"Whether it's loaded or not loaded?"

"Absolutely, Sir."

"The accused has pointed a weapon in your general direction before, right?"

"Roger, Sir,"

"And it made you feel uncomfortable?"

"It did, Sir."

"With good cause, isn't it?"

"Yes, Sir."

"You should never point your weapon at a combatant in arms; should you?"

"This is true, Sir."

"You know the accused pled guilty to negligent homicide, don't you?"

"Roger, Sir."

"Flagged other soldiers?"

"Yes, Sir."

"He pointed this loaded weapon at Sergeant Tackett's head?"

"Sir, I was not within eyesight of that incident. I can't say…"

"Did you know that he then flipped the selector switch from safe to semi?"

"Apparently, Sir."

"Did you know that the accused, then, pulled the trigger?

"Objection, Your Honor. He's going outside of the scope of direct examination, again."

"Your Honor, defense counsel went into the night of 23 June, Your Honor."

"Afterwards."

"You went into the aftereffects. I'm going to allow him to see what he knew about the incident itself."

"Again, did you know that the accused flipped the selector switch from safe to semi?"

"Sir, I can't say for sure that I knew. I can deduce from the facts that happened that that was probably the instance, but I can't say for certain that that's what happened."

"Would it surprise you to know that the accused, then, pulled the trigger?"

"Negative, Sir."

"You understand that negligent homicide is a serious offense."

"I do, Sir."

"We've lost a combat brother in arms?"

"Absolutely, Sir."

"Not to the hands of the enemy?"

"Roger, Sir."

"But by the negligence of the accused?"

"Yes, Sir."

"Thank you."

"Do you need to defend anything further?"

"No, Your Honor."

"May this witness be permanently excused."

"Permanently excused, Your Honor."

"No objection, Your Honor."

The Defense called Sergeant Clifton Cogdell as a witness. I saw SGT. Cogdell got up, looked directly at me, and nodded his head. He was, in my opinion, the best NCO I had. It was such a privilege to work with him.

"State your name and unit for the record, please?"

"Clifton Cogdell Junior, Sergeant Cogdell."

"And what unit are you with, Sergeant Cogdell?"

"Alpha Battery, 1-76, 4th Platoon, Gun truck, lead scout."

"Do you know Lieutenant Davis?"

"Yes, I do."

"And how long have you known Lieutenant Davis?"

"I got to the unit on the 1st of December, but I didn't know him until January when we got over here."

"January 2005?"

"Yes, Ma'am."

"And how did you come to know Lieutenant Davis?"

"He's our platoon leader."

"And how long has he been your platoon leader?"

"He's been our platoon leader ever since the incident."

"Since January all the way up until...I don't know

when the incident happened."

"Around the June timeframe."

"How would you describe Lieutenant Davis's leadership style?"

"Okay, I put it like this. As an NCO, you have to tell your leaders, sometimes, some of the things that need to be done correctly. He'd listen to it, go back, and make the changes that needed to be done. He led from the front, I'd say, because he finally got up there and was the lead scout, instead of letting just one person do it all the time."

"And how did you enjoy working with Lieutenant Davis?"

"I enjoyed it. He made it fun. You go out there…he listened to me. That's what NCOs and officers do; we help each other out. I just say that…tell him some of the things that I experienced the last time I was over here; some of the things that I saw that we could do better. He listened to me, empathized. Some of the things I didn't understand from higher up, we talked about them. He explained to me why we had to do it this way, and I did it.

"From your observations, did Lieutenant Davis take care of the soldiers in his platoon?"

"Yes, Ma'am."

"And would you work with Lieutenant Davis again?"

"Yes, Ma'am. I was going to say something else, but yes, Ma'am. I was going to say, "Hell yeah."

"I'm sorry?"

"I was going to say, Hell Yeah."

"Your first answer was probably the better one."

"Have you talked to Lieutenant Davis since the 23rd of June 2005?"

"I had to take one of my soldiers to check on a bonus for reenlistment, and as we were there, I had a chance to see him."

"And what was Lieutenant Davis most concerned with when you spoke to him that day?"

"He talked about the soldiers in the platoon; how's the platoon doing; how's everybody in the battery doing? We talked about Tackett and his family. I tried not to talk about that and tried to talk about other things, but that's all he talked about, really, was what's going on with us, the battery Tackett, and everyone else, but I tried to push away from that, but it didn't work."

"Thank you, Sergeant Cogdell."

"Government."

"You've been in the military for 11 years?"

"Yes."

"Been to Basic Training?"

"Did I do Basic Training? Yes."

"AIT?"

"Yes, Sir."

"PLDC?"

"Yes, Sir."

"Muzzle awareness; important, isn't it?"

"Yes, Sir."

"Never point a loaded weapon, whether loaded or unloaded, at another soldier: right?"

"No, Sir."

"When you're in red status…you know what red status is, right?"

"Yes, Sir."

"When you're in red status, you don't point your weapon at other soldiers; right?"

"No, Sir."

"Put it from safe to semi?"

"No, Sir."

"Pull the trigger while it's pointed at somebody's head?"

You don't do that, do you?"

"No, Sir."

"Thank you."

"Defense, anything further?"

"No further questions, Your Honor."

"May this witness be permanently excused?"

"Permanently excused, Your Honor."

"No objection, Your Honor."

"Defense offers the good soldier packet, Your Honor, and requests that a copy be substituted for the record, Your Honor."

"Any objection to Defense Exhibit A for identification?"

"No, Your Honor."

"Thank you, Your Honor. Defense calls Lieutenant Davis for an unsworn statement, Your Honor."

"Lieutenant Davis, how old are you?"

"24 years old, Ma'am."

"And where did you grow up?"

"In Orange, New Jersey."

"Why did you decide to join the military?"

"My uncle, my mother's brother, was a United States Marine and my chief mentor when I was growing up. He had good observations and good thoughts about the military, not particularly the Army, but the military as a whole. And through his observations, I was led to study the military, American military history, discipline, and leadership. And from an early age, I was intrigued by it."

"How long have you been in the Army, now,

Lieutenant Davis?"

"A little bit over two years."

"And what type of assignments have you had?"

"I have been the Fire Direction Officer for Charlie Battery, 1st Battalion, 41st Field Artillery, and then carried over when we stood up Alpha Battery, 1st Battalion, 76th Field Artillery. At the Joint Readiness Training Center, I became a platoon leader and have had that position since then."

"When did you deploy to Iraq?"

"We deployed in country to Kuwait around February 1st, but I deployed in Iraq ahead of my battalion on the 8th of February 2004."

"How did you enjoy being the platoon leader here in Iraq?"

"It was good. It was really good."

"What did you like most about being a platoon leader?"

"The soldiers. Interaction with the soldiers. I have had great NCOs and even better soldiers."

"Is there anything else you would like to tell this court?"

"Yes, I do. Being an officer, you have certain duties that you have to uphold at all times. There's always a duty to lead by example and lead from the front at all times. And as an officer, I'm charged with carrying out specific duties. That night, I should not have been flagging my soldiers with my weapon. I should have paid more attention to muzzle awareness and adjusted my tactical sling so that it would stay in a position where my weapon's muzzle was always pointed to the ground. And I definitely should have corrected the bad habit of switching my safety selector switch back and forth from safe to semi.

"My actions, there's no...I'm still. There is no forgiving them. I don't see how I, as an officer, could do something so retarded and so stupid. It took life from Sergeant Tackett, and I knew Sergeant Tackett. As were all of my soldiers, he was very dear to me, because I was always very personable with, not only the soldiers in my platoon, but also the soldiers in the battery. He was a very good soldier. I just wish that his parents and his wife, and his friends could kind of forgive me, even though it's probably too much to ask, in their own time. There's no excuse for my actions, whether it's negligent or willful, at all, especially since it resulted in the death of Sergeant Tackett."

"I have to live with this for the rest of my life. There's a good part of me...a very good part of me that feels that maybe I should just go away for life. That's not going to change, no matter what happens." There is no price you could put on somebody else's life, whether it's a day or 3 years and 9 months, there's no price. No time they could ever put... so I really do ask and hope that the soldiers and all soldiers, everybody that's here, and the parents and friends of SPC Tackett, that someway and somehow, you could find it in your heart to forgive me for my negligence, even though I probably don't deserve it. It'd make me feel a little bit better. That's all."

"Thank you."

"Defense rests, Your Honor."

"Government, do you have any rebuttal evidence?"

"On 23 June 2005, 22-year-old Joseph Tackett was shot and killed. He was the first casualty of 7th Platoon, Alpha Battery, 1st Battalion, 76th Field Artillery Regiment." Joseph Tackett was not on patrol when he was killed. He was not killed by an IED. He was not killed by a sniper's bullet. He wasn't killed in an engagement with insurgents. He was killed right outside of his barracks room, while sitting and joking with his friends, waiting for their nightly platoon brief. 22-year-old Joseph Tackett was killed by a fellow American, an officer, a platoon leader. Lieutenant Willie Davis fired a bullet from his M-16, which hit Sergeant

Joseph Tackett in the head. This bullet entered the left side of his head and exited behind his right ear.

"This incident was not a simple accident. The killing of Sergeant Joseph Tackett was a tragedy waiting to happen. An accident is when something happens by chance, by some convergence of events that an individual cannot control. This was no accident. Lieutenant Davis acted in such a consistently reckless manner that a tragedy like this was inevitable. The evidence demonstrates that Lieutenant Davis had a habit of reckless weapons handling, whether it was carrying his weapon with the barrel pointed out at soldiers or flagging soldiers for laughs, or even charging his weapon in a dining facility. Lieutenant Davis treated his weapon like a toy, not a deadly instrument of war. This was a pattern that led to a tragedy. Lieutenant Davis was not some private out of AIT. This was not a weapon with which he had no experience. Rather, he was a lieutenant…a first lieutenant in the United States Army, who had trained with this weapon for some time. He knew you should never ever point a weapon at somebody. He knew you should never flip the selector switch. And he knew you should never ever pull that trigger unless you mean to kill somebody. Lieutenant Davis was an officer. He's supposed to be an example to people. He was a platoon leader solely responsible for about 20 soldiers and countless civilians who depended on him for convoy security. We've heard testimony that Lieutenant Davis is a great officer, widely respected by his soldiers; that he didn't mean for this to happen. That may be true, but the fact of the matter is this was a tragedy waiting to happen. He had been told time and again not to flag soldiers and exercise muzzle awareness. Yet, he continued to do just that. That was a tragedy waiting to happen.

Joseph Tackett was born in September 1982 in a small town in Kentucky. He was at community college when the terrorist attacks of September 11th took place. Like many other Americans, Joseph Tackett felt a duty and answered the call. Joseph Tackett was a 22-year-old specialist promotable. As a soldier, he was second to none. He was selected as a team leader and gun truck leader when his rank dictated otherwise. This is not an assignment that was handed out lightly. Joseph Tackett was given this assignment because his chain of command trusted him to lead and to make decisions. As Sergeant First Class Karns and Specialist Allen told you, if there was a job to be done, Sergeant Joseph Tackett would get it done. Although he was a specialist, Sergeant Tackett was a leader. He was even…but in this leadership position, he still maintained the traits of camaraderie and his humor and constantly would pick spirits up, and he remained a very,

very popular soldier. He was a genuinely nice guy whom many were proud to call a friend.

The Army was not meant to be a permanent career path for Sergeant Tackett, but that didn't mean he was skating just waiting for the exit. No, he continued to work as hard as he could and got the job done. That's why, as a specialist, he was given his own truck. He acted like a soldier was expected.

Sergeant Tackett did not know what his life entailed. We heard that he wanted to serve a Jägermeister McFlurry. What 22-year-old Joseph Tackett could never have known was that he would die at the hands of another American. An officer, a platoon leader. This was a tragedy waiting to happen.

7th platoon, Alpha Battery, 1st Battalion, 76th Field Artillery Regiment is responsible for convoy security. It is a vital mission to ensure Iraqi and American civilians get to where they're supposed to and they get there safely. For four whole days, this platoon was taken off the line. And it was only when the platoon leadership, the platoon sergeant, basically said, "Hey, we need to get their minds off this tragedy," that they were sent back on the line. Now, I am sure that some of them still had it in the back of their mind, and yes, they were out on patrol, but it had...it was still in the back of their mind.

Lieutenant Davis was reckless and careless with the treatment of his weapon. He had a history of flagging other soldiers. Soldiers would complain, and he would ignore it. He would charge his weapon in the DFAC and laugh at it. Had Lieutenant Davis exercised the proper level of care and concern, 22-year-old Joseph Tackett would still be alive. Instead, through his reckless behavior, he set the stage for this. A tragedy waiting to happen. All Lieutenant Davis had to do was properly clear his weapon, and 22-year-old Joseph Tackett would still be alive. All he had to do was not point his weapon at anybody, and Mrs. Tackett would still have a husband. All Lieutenant Davis had to do was not switch that selector switch, and his parents would not be mourning the loss of their 22-year-old son. Had Lieutenant Davis not pulled that trigger that fateful night, 22-year-old Joseph would still have a bright future ahead of him. In light of the facts and the need to show that muzzle awareness is a serious matter, the government asks for a sentence of 40 months confinement, a dismissal, and total forfeiture. Thank you, Your Honor."

"Defense."

"Yes, Your Honor."

"Your Honor, the prosecution is right. This is, indeed,

a tragedy. Christopher Frye, an English playwright, once wrote, "That, in tragedy, every moment is eternity." The night of 23rd June 2005 is indeed a tragedy and a moment that will live in Lieutenant Davis's memory for eternity. Sergeant Tackett's life was tragically cut short, but there were two lives that we lost that night, as one of the government witnesses pointed out to you: that of Sergeant Tackett and that of Lieutenant Davis, as well.

Lieutenant Davis made a grave and very tragic error in judgment that day. And even though he is alive and sits before you today, that is an error, a tragic error that will haunt him for the rest of his life. As you've heard from the witnesses and the statements that you have before you, Lieutenant Davis has…had a very promising future, which is now lost. He's only 24 years old and has been in the Army for just over two years. And these are the comments that have been made about Lieutenant Davis: a quality junior officer, a superb young officer with unlimited potential, who truly cares about his soldiers, goes out of his way to take care of his soldiers, dependable, reliable, responsible, and loyal to the mission.

As a person, there's also a lot of good in Lieutenant Davis, as you have heard, and as the statements you have before you show you. And these are the comments that have been made about Lieutenant Davis; he may be a little shy and reserved, as you have seen yourself, today, but he is a true and caring friend; a caring young man, has a big heart and will do all he could to make sure those around him were safe and happy, great friend and confidant; compassionate, loyal, respectful, and loving.

Your Honor, Lieutenant Davis is not asking you to discount the life of Sergeant Tackett. We're simply asking that you consider the life that has yet to be lived by Lieutenant Davis at age 24. Thank you, Your Honor."

"Thank you, Counsel, the court is closed for deliberation."

9

THE VERDICT

Peter 4:16:

"Let him not be ashamed, but let him glorify God in this matter."

When His Honor left the courtroom to deliberate on my case, I was finally able to gather myself, just a little. One by one, the soldiers filed out into the hallway between the courtroom and the judge's chambers. I remained seated, frozen in place, trying to process everything that had just transpired over the last few hours. The weight of it pressed hard against my chest. I knew confinement was inevitable. I just didn't know how long. Still, I believed it was only fair. I deserved to serve time, even though I was scared out of my mind. That didn't matter. What mattered was that my recklessness and carelessness had cost a man his life. If I had just shown a shred of caution, a bit of discipline, SPC Tackett would still be alive.

After sitting in silence, I eventually forced myself to stand and step into the hallway. That's when I saw PFC Mattis. He was crying. Quietly. Uncontrollably. I approached him, placing a hand on his shoulder.

"What's wrong, Matt?" I asked, though I already knew the answer.

"Nothing is wrong, Sir," he said, eyes downcast, voice broken. "I just… I just can't believe all this happened."

"No one wishes it hadn't happened more than I do, Matt," I said softly. "I'm so sorry for everything. For all of it."

While Mattis sobbed beside me, I caught a glimpse of my battalion commander and the new Alpha Battery commander standing off in the distance. The look on his face was hard, perturbed, maybe even disgusted. I couldn't bring myself to face him. My stomach began to twist into knots, the same kind of knots I felt that night in the psychiatric hospital before the medication dulled the pain enough to sleep. The memory of SPC Tackett hitting the ground after I fired that shot surged through me again like a tidal wave. I could hear the prosecutor's voice in my head: reckless… careless. And he was right. I had been all those things and worse.

I had taken a man's life. Stolen his future. Taken him from his wife, from his family. The toll of that truth was unbearable. I turned away, unable to face anyone else, and returned to the courtroom. I sat back down in my seat, despondent, heavy with shame.

Moments later, CPT Ku reentered, followed by the others. Their presence signaled one thing: the judge had made his decision.

"All rise," the Sergeant at Arms commanded as Colonel Reinert entered the courtroom. The room stiffened with formality and tension as he walked up the center aisle, his boots striking the floor in deliberate rhythm, and took his place on the bench.

"Court is called to order. The record should reflect that all parties present when I recessed to deliberate are again present. Defense counsel and accused, please rise."

I stood as ordered, locking my body into the position of attention even as my heart pounded inside my chest. My legs felt weak, my palms damp. I stared straight ahead, bracing for the words that would shape the course of my future.

"First Lieutenant Willie Davis Jr.," the Colonel began, his voice clear and firm, "this court-martial sentences you: To forfeit $2,500 of your pay per month for two years and six months; to be confined for two years and six months; and to be dismissed from the service of the United States Army."

The words rang in my ears, echoing louder than any gunfire I had ever heard in combat. Two years and six months. That was the sentence for my recklessness. For my failure. For my carelessness that had taken SPC Tackett's life. My stomach tightened, and I fought to stay upright, to keep from breaking down.

Thirty months.

It wasn't enough. It could never be enough.

What sentence could ever repay a life?

"Court is adjourned," I heard, barely registering the words as they pierced the fog of guilt clouding my mind.

CPT Ku turned toward me and whispered gently, "Well…It's only thirty months, Lieutenant Davis. Everything's going to be all right."

I nodded weakly, struggling to find my voice. "Can you let my wife know what my sentence is, Ma'am?"

"I will," she said, placing a comforting hand on my shoulder. "Take care of yourself, Lt. Davis."

"Thank you, Ma'am."

10
CONFINEMENT

Psalm 69:33:

For the Lord hears the poor, and despiseth not his prisoners.

I arrived at the Regional Correctional Facility at Fort Knox, Kentucky, on September 23, 2005, alongside a group of other convicted soldiers. "God help me," I whispered aloud, heart pounding as I stepped toward the looming entrance. It felt like I was walking into a tomb, and in many ways, I was. This was the burial of the life I once knew.

My hands trembled as I walked, fear coiling tightly in my chest. I didn't know what to expect. I didn't know when I'd see my wife, my mother, or my father again. The cold, mechanical doors opened automatically as I approached, and I stepped through them with the weight of the world pressing down on my shoulders.

"Face the wall and get at parade rest!" barked a guard, jolting me back into military instinct. I snapped to attention, old habits overriding my new reality.

First Sergeant Connelly, our escort, announced our arrival: "Checking in prisoners from Camp Arifjan, Kuwait." Just like that, it was official. I was no longer a soldier. I was an inmate.

Inside, we were frisked and led into a bathroom for a strip search. "Drop your clothes," a young private ordered. Humiliated, I complied. A civilian cadre stepped forward and addressed us with cold authority: "You're no longer soldiers. You're inmates now. From this point forward, you follow the MGI, the Manual for the Guidance of Inmates. This is your Bible. This is your rule book."

Every strip search chipped away at my dignity. I felt like less than a man, less than human. Each time it happened, I cringed inside. Each time, I whispered the same prayer: "Lord, do I really deserve this?"

After processing, we were issued brown uniforms, a black footlocker, basic hygiene items, a small notepad, and ten envelopes. Then came the next order: "Inmates, pick up your shit and proceed to General Population." We moved silently, watched closely by guards and surveillance cameras. There was no room for individuality, only compliance.

"South Wing, Control place inmate Davis in Kilo Cell," a cadre called out.

I walked down a long corridor flanked by steel and silence. Two wings split the facility, the West Wing for new and medium custody inmates, and the South Wing for minimum custody. A bench waited outside the tinted glass of the control room, where monitors watched our every move.

Finally, I was led to a narrow seven-by-four-foot cell. A camera stared down from above. The iron bars slammed behind me, their echo sealing my fate.

I dropped my belongings on the floor, turned toward the bars, and gripped them tightly. I looked around that dim, suffocating space. And then the weight hit me all at once.

I broke down.

Tears poured from my eyes. This was real, my military career was over. My identity as a soldier was gone. I was no longer Lieutenant Davis, the proud platoon leader of 4th Platoon, Alpha Battery. I was Inmate Davis. Convicted. Stripped. Forgotten. Two and a half years in confinement. Thirty months to think about everything I had lost.

As days turned into weeks and weeks into months, I learned the only way to survive was to stay busy. Older, more seasoned inmates gave the same advice: "Keep moving, keep working, keep your mind occupied." And so, I did.

I worked during the day. Afterward, I went to the facility library and checked out every book I could get my hands on. I read until my eyes were sore. I exercised each night until my body ached. Anything to drown out the regret, the pain, the memories.

And when that wasn't enough, I reached for the phone.

Each inmate was allowed to list up to ten contacts. We could speak for twenty minutes per call, as long as we had funds in our account. My wife, Kristen, God bless her, always made sure there was money in mine. Month after month, she sent me what I needed so I could hear her voice. Talk to her. Talk to family. Talk to friends. It was those conversations that reminded me I still had a reason to hope. A reason to hold on.

However, what brought me the most strength was something unexpected. I read the Bible while in confinement, and it became my source of strength, of clarity, of something I couldn't name at first but would later recognize as hope. In a place where time seemed frozen, where the silence was heavy and the walls indifferent, I found a voice that spoke directly to my soul.

Confinement strips away the noise. There are no distractions, no illusions of control. Just you, your thoughts, your regrets—and if you're fortunate enough to reach for it—God. At first, I opened the Bible out of routine, maybe even guilt. But over time, it became more than that. I wasn't just reading Scripture; I was being read by it. Page after page, it exposed me—my pride, my fear, my failure—but it didn't leave me there. It offered grace.

Verses I had skimmed over before took on new weight. Isaiah, Psalms, Romans, and books I had once treated as doctrine became lifelines. I found comfort in Paul's reminders of redemption, in the quiet assurance that I was not beyond forgiveness. But what really resonated with me was King David's cries for mercy. His words in the Psalms were not just ancient poetry, they were my own pleas, echoing from a place of guilt, regret, and longing to be made whole. Like David, I failed. And like David, I threw myself at the mercy of God, not because I deserved it, but because I had nowhere else to go. His prayers were also not polished; they were raw, desperate, and honest. David's cry in Psalm 51— *"Have mercy on me, O God, according to your unfailing love"*—became a lifeline. I whispered it in the dark. I prayed it in silence. I let it say what I couldn't yet articulate: that I was broken, that I needed mercy, and that somehow, I still believed it could be given. David begged God to blot out his transgressions, to wash him clean, to create in him a pure heart. I read those words over and over and over again. I wasn't just reading David's story—I was living it. I, too, had blood on my hands. I, too, knew the weight of failure. And yet, in David's brokenness, I saw a path forward. Not one that erased the past, but one that led through repentance to restoration. At times, when the gate slammed shut behind me, I held on to the quiet conviction that the same God who saw Moses in the desert and Paul in prison saw me in my cell.

There were days I couldn't pray, only read. Days I couldn't speak to anyone but somehow felt seen. That's what the Word became in confinement—not just a text, but a presence. A reminder that even in a cage, even at the lowest point of my life, I was not abandoned.

I used to think of strength as something proven in combat or command. But true strength, I learned, is found in surrender, in brokenness. In letting go of everything you thought made you worthy and realizing that the only thing that truly sustains you is grace.

But it would be a long time before I could fully come to grips with what had happened—what I had done. As the days wore on, I began to wrestle with myself—my mind, my spirit—trying to make sense of a reality I could neither undo nor escape.

I was confined. But through the Word of God, I began to heal.

Yet healing did not come all at once. It unfolded in stages over years—through waves of guilt, layers of shame, and moments of painful revelation. The same Word that comforted me also confronted me. It held up a mirror to everything I had become, and everything I had failed to be. And in that mirror, I saw not only my sin of negligence, but the possibility of redemption.

PART III:
WRESTLING
WITH FAITH

11

THOUGHTS: "IT WAS ALL MY FAULT"

Romans 8:18:

"For I consider that the sufferings of this present time are not worth comparing with the glory that is going to be revealed to us."

"I do not deserve to live." In my mind and at this time in my life, I was as close to being physically and mentally suicidal as I was at any time during my efforts to rehabilitate myself while I was in confinement at Fort Knox. How can a man take a life, accidental or not, and live with himself?" This is the question that has haunted me since the dreadful moment, "Lord, how can I live with myself, how can I get through this. Lord help me, please. I can never get any sleep. "God, why did this happen?" "It was all my fault." "Lord, tell me what I need to do?"

While at Fort Knox, I received "therapeutic rehabilitation" from the social workers' department at Fort Knox. The department was led by Mr. Steve Bowen and Mrs. Sarah Wellman. Mr. Bowen was an elderly gentleman who I felt had a good heart and a good outlook on my situation. One day, Mr. Bowen called me into his office to introduce himself as my primary social worker. "Inmate Davis," he said as I approached his office door. "How are you doing, Sir, today? I understand that you wanted to see me." I said. Mr. Bowen was a man who was short in stature with a very heavy, stylish, grayed beard.

"Yes, I did call for you. Please sit down," he said as he motioned to the leather brown chair in front of his desk.

"How are you doing today?" he asked,

"How are you getting along with the Cadre and the other inmates?"

"I am getting along fine, I guess. I am not sleeping very much, but I am fine," I replied.

"Well, my name is Mr. Steven Bowen, and I will be your primary social worker here while you are confined here at Fort Knox."

"Ms. Wellman," he continued, "Is the facility's assistant social worker, and if you wish to meet with us, just drop a DA Form 510 and one of us will call for a meeting with you within a week."

A DA Form 510 was the facility's way of granting its inmates certain requests. While confined at the facility, I had to drop a 510 in order to receive anything, including visitation rights, the reissue of clothing, how many phone numbers I wanted on my outside calling account so I could talk to my family and friends, and even meetings with the facility chaplain.

Mr. Bowen went on to further explain his service and his job at the facility, as well as the goals of the social work department at the Regional Correctional Facility. "Inmate Davis, I want you to know that we are committed to the rehabilitation of our inmates here at the RCF."

"Sir, what type of rehabilitation are you referring to? In my case, respectfully, I have to say that there is no rehabilitation in my situation," I said.

I paused for a minute as I saw him looking at me intently as if he was taking in my every word.

"Mr. Bowen," I continued.

"I have to deal with what happened for the rest of my life. I do not think that there will be a day in which I will not think about what happened that evening in which I shot one of my soldiers negligently in the head during this God-awful war," I said.

I have these dreams, and all day long, I think about SPC Tackett and his family and how they are feeling, and it does nothing but tear me apart inside and out.

"Tell me, Sir, how can I be rehabilitated? How can you help me?" As he took in what I just said to him, I was anxiously awaiting his reply. After he leaned back in his chair, he looked at me and said, "In your case, we cannot rehabilitate you." "You have just been ostracized from society and sentenced to confinement for up to thirty months at this military confinement facility."

"You are not a criminal, you are not a sexual offender, or a thief, or anything else. You were essentially negligent and your duties as a soldier and an officer in which resulted in the taking of the life of another soldier." He stated.

"I can only imagine how you may feel." "How are you feeling?" he asked gently.

I paused. At times, I didn't even know how to answer that question. My emotions were a tangled mess, a heavy, suffocating conglomerate of feelings I had never experienced before.

Most days, I hated myself. I hated how careless I had been, how I failed to check and double-check as I should have. The guilt was relentless, clawing at me every waking moment. Since entering confinement at Fort Knox, dark thoughts began to creep into my mind—thoughts I never imagined I was capable of having.

I started to wonder if the only way to bring relief and justice to SPC Tackett's family was to take my own life. Maybe, I thought, if I killed myself, it would somehow balance the scales, somehow ease their pain.

Before all of this, suicide had never even crossed my mind. It was unimaginable, the furthest thing from anything I thought I was capable of considering. But now... everything felt different. Everything had changed.

At first, I didn't want to tell Mr. Bowen how I was really doing. I wasn't ready. Maybe it was because I didn't yet feel comfortable with him, or maybe I was still trying to keep parts of my pain buried deep. I had only just begun to warm up to LTC Breitenbach, allowing myself to open up slightly, and then suddenly—after my court-martial—I never saw her again. I was moved so quickly into military confinement that it felt like the fragile progress I had made was ripped away.

After that initial meeting with Mr. Bowen, I didn't see him again for about a week. In the meantime, I began to settle into my temporary—but not temporary enough—life in confinement.

I was informed that although I had been sentenced to 30 months, five months had already been deducted for good behavior, provided I stayed out of trouble while confined. That meant I was now looking at a total of 25 months— though there was still a possibility of earning additional time off by participating in work details and securing a job within the facility.

The Cadre assigned me to work in the dining facility. Alongside other inmates, I spent my days washing dishes, wiping down tables, and mopping floors. The routine was rigid, almost mechanical. Each morning, we were woken at First Call at 0600. By 0615, a soldier would come by each cell to perform what they called "COUNT."

It was a precise and deliberate process. The soldier would move from cell to cell, counting every inmate in the facility. Each of us was required to stand at the position of attention in front of our bunk. When approached, we had to execute an about-face and sound off with our first and last name, followed by our bunk number.

By 0630, once the count was cleared, inmates assigned to the morning shift in the DFAC were instructed to report for duty, while the others were ordered to clean their cells and their respective areas.

I was placed in a sixteen-man cell lined with cold metal bunks, each outfitted with a thin, two-inch green mattress and a single pillow. In the center of the cell was a television mounted high on the wall and a four-man stainless steel table bolted firmly to the ground. The shared space also included two showers, two toilets, and two sinks—all offering little privacy in an already crowded and uncomfortable environment.

The company I was forced to keep was, as one might imagine, far from compelling. I was now a convicted felon, surrounded by what I considered *real* felons. The other inmates in my cell were there for a wide range of offenses: some for adultery, others for various types of larceny, and a few for aggravated assault. But what struck me most and what felt especially demeaning was being confined alongside rapists and child molesters, men from nearly every division of the active-duty Army, National Guard, and Reserves.

I couldn't shake the feeling that I didn't belong there. I was the only one whose crime was the result of negligence, something unintentional. Yet, no matter how I tried to rationalize it, I couldn't escape the truth: I was one of them now. A felon.

The weight of that reality was something I had to come to terms with alone. No one else could understand the thoughts racing through my mind, the enormous guilt that consumed me, the crushing shame I believed I had brought upon my wife and my parents, or the ignominy I felt I had cast over the service and the officer corps.

Each day as I went to my assigned place of work in the dining facility, my mind drifted to the Tackett family. I couldn't stop wondering what they were going through, what they were feeling, and how they were coping with their loss. How were his mother and father holding up? How was his wife managing to get through each day, carrying the unimaginable weight of knowing her husband was killed in Iraq—killed as the result of a negligent fratricide committed by one of his own officers?

I often thought about my platoon and how they were faring in Iraq. Was Private Kieffer still struggling to drive for his section chief? Was Sergeant Cogdell employing the right tactics and procedures as they navigated the treacherous streets of Baghdad? And what about their new platoon leader—how was he doing leading my men? *Was he taking care of them? Was he listening to my NCOs and leading by example?*

My thoughts were many, but above all, I found myself wondering how I could ever make this right—with God, with SPC Tackett's family, and with myself. In one way or another, they were always on my mind.

As I began serving my sentence in confinement, I learned quickly that each new day brought its own challenges. As I expected, the Cadre kept us inmates busy, following a rigid and highly structured schedule. Every weekday morning, we were woken at First Call at 0600 and

0700 on weekends and holidays. We were expected to rise immediately, make our beds to precise standards, and clean ourselves up before the 0615 morning count.

"Count," as they called it, was a critical procedure the Cadre conducted at least four times a day to ensure every inmate was present and accounted for. Life inside was as structured and controlled as any I had known in the Army. After the first morning count, we were marched to breakfast—wing by wing, cell by cell. I soon learned that each wing consisted of six sixteen-man cells, and the Cadre frequently rotated us between cells to break up cliques and minimize the potential for physical altercations.

After about a month and a half in confinement, I began to consciously settle into the realities of my situation. My wife, Kristen, became my lifeline—her love and support gave me hope, even though we were miles apart and separated by the constraints of the law. In an incredible act of devotion, she moved from Atlanta to Louisville in just three weeks so she could be closer and stand in support of her husband.

I was deeply grateful for her—more than words could ever convey. And yet, deep down, a part of me wished she didn't have to bear the weight of this ordeal. I couldn't begin to imagine the storm of emotions she must have been navigating, but in my heart, I knew it had to be unbearably hard for her.

She came to visit me almost every week after she arrived in the Louisville metropolitan area. The facility allowed visitation three days a week, two hours on Wednesdays, and nearly all day on Saturdays and Sundays. Most times, one of the Cadre would approach my cell and call out, "Davis, report to Post One. You have a visitor."

Before leaving, they would frisk me to make sure I wasn't taking anything out of my cell to give to her, then escort me to the visitation room. At the entrance, I'd hand over my prisoner identification badge to the guard, take a deep breath, and walk inside. Another guard kept a watchful eye on the room at all times, carefully observing every interaction between inmates and visitors.

"Hey, you, how are you doing?" she'd ask the moment she saw me. Every time, I forced a smile and said, "I'm doing well." But deep down, I knew I was lying through my teeth. I wanted to seem strong for her, to protect her from the weight I was carrying. But we both knew the truth: I wasn't okay. Not even close.

At my seat, I lingered just long enough to share a quiet, meaningful embrace, feeling the comfort of her presence. In those moments, God, I felt so blessed—so full of joy. I was one of the few inmates in the facility who even had a visitor.

"Now you know I can't go a week without seeing my Will!" she said emphatically, her brown eyes locking with mine. And every time I looked into them, the walls of the confinement facility seemed to fade away. For those few precious moments, I didn't feel so alone. For a while, I could almost believe that everything was going to be okay.

"So... how are you really feeling, Will?" she asked gently. She knew me too well to accept my usual surface-level answers.

"I'm okay," I said, my voice low. Then, unable to hold back, I muttered, "I just wish this had never happened." My voice trembled as anger began to boil to the surface. "How could this happen, Kristen? How could I be so fucking careless?"

She reached across the table, her soft hands gently rubbing my head as she whispered, "It's going to be alright. You have to forgive yourself, Will."

But I couldn't. I wasn't sure if I ever could.

I can't bear to live with myself, I thought. I wasn't sleeping well. Dark thoughts of suicide haunted me night after night, convincing me that I had no reason to keep going. Kristen gave me plenty of reasons for one reason in herself, but I chose to ignore them. I felt I didn't deserve her, or any happiness for that matter.

That same month, to my dismay, I learned that news of SPC Tackett's death had traveled stateside.

A few weeks earlier, shortly after arriving at Fort Knox, I had been summoned to the office of the Chief of the Prisoner Services Branch. The office was led by a stern Master Sergeant whose presence seemed to fill the room. As I stood at parade rest, bracing myself for whatever news was coming, I saw the imposing figure of MSG Lamoth step forward.

"Inmate Davis," he said, his tone firm and measured. "The Courier-Journal—the local newspaper here in Louisville—contacted me this morning," he stated

"They are requesting a phone interview with you regarding the actions that took place on 23 June 2005, which led to your confinement here at Fort Knox," he continued.

His words hit me like a blow to the chest.

I felt the blood drain from my face as the reality of my situation sank in deeper. The shame and dread I had carried since my court-martial now seemed to crash over me in full force. I had known, deep down, that eventually my negligent actions would come back to haunt me—not just in the courtroom or on the confines of Fort Knox, but in the eyes of the American public. I would have to answer for how a United States Army officer, charged with leading American men and women both at home and on the battlefield, had accidentally killed one of his own soldiers.

"No, Cadre," I said finally, my voice trembling slightly. "I do not wish to conduct an interview with the local newspaper at this time."

I could feel my body physically shaking as I spoke. MSG Lamoth studied me for a moment, his eyes narrowing slightly as if he could see the fear coursing through me.

"Alright," he said at last. "Then you are dismissed. Report back to your holding cell, Inmate."

As I walked back down the long, sterile hallway to my cell, my mind spiraled. The weight of embarrassment I had imposed on myself now threatened to extend beyond me—onto my family. I knew the time was drawing near. Sooner or later, the whole country would know.

Two weeks later, Mr. Bowen called me into his office.

"Inmate Davis, come in," he said, his eyes scanning the papers scattered across his desk as though searching for something. "Have a seat. How are you doing?"

"I'm doing okay, Sir," I replied, trying to sound steady. "Just taking it one day at a time, you know."

"Good, good," he said absentmindedly as I sat down. Then he reached across his desk, pulling a folded newspaper toward me.

"Well, I have something I need you to look at," he said, sliding it across the desk.

I hesitated before taking it in my hands. It was the October 11, 2005, issue of *The Louisville Courier-Journal.* Bold black letters screamed across the front page:

"Soldier Dies a Negligent Death: Kentuckian Shot by Officer in Iraq."

My heart plummeted.

Below the headline was a photograph of a grieving woman, her face buried in her hands. I didn't have to ask— I knew it was SPC Tackett's mother.

Leaning back in the chair, I felt the weight of the moment press down on me as I began to read the article written by Alan Maimon:

WHITEHOUSE, Ky. — When Wendell and Kathy Tackett learned their son had been shot and killed in Iraq on June 23, they assumed enemy fire had hit him. But SPC Joseph Tackett was killed by a U.S. soldier—one of 10 homicides since Operation Iraqi Freedom began in March 2003. Even more rare, his death was at the hands of an officer, the military said.

The Pentagon hasn't released an official report on the death, but military officials said the 22-year-old was killed by a lieutenant who pointed his M-16 rifle at Tackett in a "safe haven," a place where loaded weapons are forbidden.

"He was killed by a stupid, senseless, irresponsible act," said Kathy Tackett, 52, a food-service manager.

At a court-martial, Lt. Willie Davis pleaded guilty Aug. 31 to negligent homicide and negligent dereliction for failing to clear his weapon and maintain muzzle awareness, said Lt. Col. Clifford Kent, spokesman for the Army's 3rd Infantry Division at Fort Stewart, Ga.

The maximum sentence for the crime was 45 months in prison. Davis, of Lithonia, Ga., was sentenced to 30 months at a military prison at Fort Knox and discharged from the Army, Kent said.

The Tackett's, from Johnson County, said Davis should have received the maximum sentence.

"To me, that's not enough punishment," said Wendell Tackett, 53, a house builder. "But what I want most from him is an apology."

Wendell Tackett said he and his wife only recently learned Davis is being held in Kentucky and have been going through military channels to try to talk to him.

Davis declined an interview request through Gini Sinclair, a Fort Knox spokeswoman. Davis' family could not be reached for comment.

In an e-mail to the Tacketts obtained by The Courier-Journal, Col. Daniel Pinnell, Tackett's battalion commander, said he and witnesses to the shooting didn't suspect "any malicious intent" on Davis' part.

Rare case

The vast majority of deaths in Iraq—more than 1,900—have involved soldiers killed in action. That figure includes friendly-fire incidents.

As I finished reading, my hands trembled. I placed the paper carefully back on Bowen's desk, though it felt heavier now—as if the weight of its words alone could crush me. My mind spun, caught between disbelief and the cold, inescapable reality of what I had done. The shame pressed into my chest like a lead weight, making it hard to breathe.

The photograph of Kathy Tackett was seared into my memory—her face buried in her hands, the very picture of anguish. And her husband's words echoed relentlessly in my mind:

"What I want most from him is an apology."

As I sat there in silence, my eyes drifted back to the article, catching on to the chilling statistics:

197

One soldier died as a result of homicide in the 1991 Persian Gulf War out of 382 total fatalities, according to an analysis by GlobalSecurity.org.

During the Vietnam War, which saw more than 58,000 American deaths, 944 soldiers were killed by "accidental homicide" and 234 by "intentional homicide," according to the National Archives.

It was sobering to see my own failure framed within the history of other wars. And yet, none of those numbers could numb the ache in my soul. SPC Tackett wasn't a statistic—he was a son, a husband, a soldier with a future that I had cut short.

The article went on to describe how Tackett's death occurred inside Baghdad's fortified Green Zone, in the basement of one of Saddam Hussein's former palaces that had been repurposed as living quarters. Fort Stewart spokesman Rich Olson called the death "inexcusable," saying it was the kind of incident that prompts military officials to ask themselves, *"How the hell could this happen?"*

I wanted to ask myself the same question.

How the hell could this happen?

In an email to the Tackett's, Col. Pinnell had explained: "Joseph was shot during a nightly briefing with his and Davis' platoon. He was unconscious from the moment he was struck until he passed away a short time later at the hospital here in the International Zone."

The words seemed to burn into my chest.

Kathy Tackett had been told by military officials that the incident happened in a place where soldiers were supposed to feel safe—a "safe haven." Yet because of me, it became anything but safe for Joseph Tackett.

I sat there in Mr. Bowen's office, my eyes stinging and my throat tight, knowing that no amount of remorse would bring him back. And though I didn't yet have the courage to admit it out loud, deep down I knew they deserved more than my silence. They deserved an apology.

"They said Lt. Davis came in carrying his weapon and willingly pulled the trigger on a gun he thought was empty," Kathy Tackett said in the article. Other soldiers reportedly questioned why I had brought a weapon into the briefing, a clear violation of the rules. To prove it was unloaded, I pointed the rifle at SPC Tackett and pulled the trigger. That single, careless act ended in tragedy. The Army's casualty report stated plainly: *SPC Joseph Tackett died of a gunshot wound to the head.*

As I read those words, I felt my stomach turn.

The article also included a July letter Col. Pinnell had

sent to battalion spouses and family members, praising the unit's high level of training and discipline:

"No other unit in Iraq even comes close to the level of safe and effective combat patrol operations achieved by your soldiers over the last six months," Pinnell wrote. *"Our combination of intense and detailed preparation and patience in execution has made us a tough target to hit."*

The irony of his words was not lost on me. Despite all of that preparation, despite all of that discipline, I had failed in the most fundamental way—and Joseph Tackett had paid the ultimate price.

The Tackett's, still desperate for answers, voiced their anguish.

"Not knowing exactly how this happened is the worst part," Wendell Tackett said. He wanted to understand why my weapon was loaded, why the safety was off, and why I hadn't chosen to fire harmlessly into the ceiling instead.

I didn't have the words to explain it—not to him, not to myself.

Adding to their grief was a painful division within their family. Kathy Tackett revealed that her daughter-in-law, Stephanie Tackett, had refused to share information about my court-martial, including depositions from soldiers who witnessed the incident. Kathy described her son and

Stephanie's young marriage as troubled, adding another layer of heartbreak to an already tragic story.

When reached by phone at her home at Fort Stewart, Stephanie declined to comment.

The *Courier-Journal*, in a Sept. 22 open records request, had asked for copies of witness depositions and other documents related to my court-martial, as well as a preliminary report on SPC Tackett's death. The government had 20 business days to respond, and the request was still pending.

The article went on to describe Joseph Tackett's life in vivid detail.

Tackett, a 2000 graduate of Johnson County Central High School, felt a sense of duty to enlist after the terrorist attacks of Sept. 11, 2001, his mother said. He had already deployed to Afghanistan and Iraq in 2003 and, after more than a year stationed at Fort Stewart, had deployed again in January.

Kathy Tackett said her son befriended Iraqi college students, introduced them to American rock music, and exchanged emails with them while home between tours. He had even been working to complete his associate degree in general studies through correspondence courses during his deployment.

The Army had posthumously promoted him to sergeant and awarded him the Bronze Star for "meritorious valor."

Reading on, I saw that his father, Wendell Tackett, said the loss had soured him on the military. *"It's not changed me on the war, but I have hard feelings toward the government,"* he said. *"I can't be against the Iraq war because Joe knew what he was doing. ... He knew it was for the people."*

As I took in every word of the article, my eyes locked on the photograph of Kathy Tackett, her face buried in her hands, her body trembling with grief. The photographer had captured her pain in a way no words ever could.

There are no words to describe how I felt at that moment.

I looked up at Mr. Bowen and said, my voice breaking, "Mr. Bowen, I wish I could somehow console Mrs. Tackett. I wish I could sit down and explain to her—and to SPC Tackett's father—just how deeply sorry I am. But I know, Sir, that nothing I say or do can take away even a fraction of their hurt. My carelessness robbed them of their son."

For a long moment, I just sat there in his office, tears streaming freely as I whispered apologies into the air—

apologies they couldn't hear, apologies I wasn't sure would ever be enough. I could only imagine how many more lives my mistake had rippled through; how many people were carrying a piece of the pain I had caused.

"Mr. Bowen… I really want to die right now," I said finally, my voice small and raw. "I can't live with this, Sir. I don't think I can survive this guilt."

He watched me closely but remained calm. "Take a moment to compose yourself," he said softly from behind his desk.

"Sir… it was all my fault," I choked out. "I should have cleared my weapon that day. I should have executed proper muzzle awareness. I should have adjusted my tactical sling and ensured that the muzzle was pointed down, like I was trained to do. I knew the regulations. I was taught this in ROTC. How could I let this happen?"

I rubbed my eyes, trying to blink away the tears, but they kept coming.

"I read in the article that Mrs. Tackett said I *'willingly pulled the trigger on a gun I thought was empty.'* Sir, I swear to you, I never willingly pulled that trigger. Jesus Christ, I would never do such a thing. I felt the selector switch under my thumb, thought it was *safe*, and when I went to move it back, I didn't realize my finger was still on the trigger. I wasn't paying attention. God help me… I failed to do my duty, and now a man is dead because of me!"

My hands trembled as I buried my face in them. "Oh my God… I feel so sorry for Mr. and Mrs. Tackett. I wish there were something I could do to mend their pain. But even an apology feels so hollow. It's not enough. It will never be enough."

I took a shaky breath. "This should have never happened, Mr. Bowen… this should have never happened."

"Inmate Davis, do I need to move you to segregation and place you on suicide watch?" Mr. Bowen asked carefully.

"No, Sir," I said after a pause, my voice weak. "Please… just give me a moment to calm down. The last thing I need is more attention from the Cadre."

After a few moments, Mr. Bowen deemed me fit to return to my cell. A part of me knew it might have been better to go into segregation, where I could be alone with my thoughts and try to make sense of the situation. But deep down, I also understood that if I went there, the Cadre wouldn't allow me access to the phone to call my wife or my friends—those few people who, even from a distance, were helping me cling to my sanity and sense of self-worth.

At that moment, I desperately needed someone to

talk to—someone who knew me, who truly loved me. I knew that if I was left alone, with no outside encouragement, I would sink even deeper into the emotional hole I was already in.

So, after returning to my cell, I waited for the 1630 facility count to clear. Once the phones in the West Wing were switched on, I immediately dialed her number.

After the automated voice announced where the call was coming from and asked if she would accept it, I heard her faint voice on the other end.

"Hello?" she said softly.

"Hey," I replied.

"How are you doing?" she stated.

"I'm fine," I said, though the words felt hollow.

There was a brief pause before she spoke again, her voice laced with concern. "What's wrong?"

"I read the article in *The Louisville Courier-Journal*," I admitted. "Mr. Bowen called me back into the social worker's office and showed it to me."

"How do you feel, Willie?" she asked gently.

"Of course I'm not feeling good," I said, my voice cracking. "I took another man's life. How do you think I feel?"

"I don't know how you feel," she replied softly. "Talk

to me, Will. Tell me what's going through your mind."

"I have to make an effort to see the Tackett family," I said, the words spilling out. "I have to do something—anything—to try to ease their pain."

"Will," she said carefully, "what if you can never ease their pain? What if an apology is just… not enough?"

"I know that, Kristen. You're right. In a situation like this, maybe an apology will never be enough. How do you even begin to say sorry for taking the life of a child—a child I was sworn, as a United States Army officer, to protect? God… this is such a fucking nightmare. I wish with everything in me that it had never happened. But it did. And now I have to live with the crushing reality that I not only took the life of another man, but in doing so, I've shattered the lives of everyone who loved him.

For the rest of my life, this will be my burden to carry. How do you make something like this right? There's no way. There's absolutely no way."

In that moment, a dark thought crept into my mind—a thought so heavy and twisted I couldn't bring myself to say it aloud. I should tell her to divorce me now… so I can take my own life as retribution for his.

In my broken reasoning, I believed that if I did that—if I ended my life—it might somehow count as a stronger apology. Maybe, just maybe, it would balance the scales for the life I had taken.

As the days turned into weeks, and the weeks slowly stretched into months, I began to come to terms with my life as it now stood.

After three months in confinement, I was elevated to minimum custody. That change meant I could begin working details outside of the facility, a welcome change. Minimum custody inmates were given the privilege of working at the Fort Knox Recycling Center, allowing them to accumulate extra days off their sentence. For me, it wasn't just about earning time off; it was about having something— anything—to focus on beyond the four walls of my cell.

Still, I couldn't shake the thought that the longer I remained in confinement, the harder it would be to move forward with my life, and even harder to attempt to make amends—if such a thing were even possible—for the anguish I had caused the Tackett family. The truth was, I still didn't know how to do that.

The article in *The Louisville Courier-Journal* continued to haunt me. Night after night, as I laid my head down, the image of SPC Tackett's mother burned in my mind. The pain etched into her face in that photograph seemed eternal, as though no measure of time could ever ease it.

And I began to accept that no matter what I said or did, there would never be a way to fully take away her suffering. But I could still offer something—a strong, heartfelt apology.

In my heart, I resolved that the best thing I could do was to live a life that itself became an apology. A life where every action, every decision, and every reason for waking up in the morning would be rooted in an effort to honor SPC Tackett and his family.

In essence, I decided to dedicate the rest of my life to them.

My time in confinement became more than just punishment; it became my preparation. Preparation for a future where, perhaps, I could build something meaningful out of the wreckage I had caused. It became the motivation I needed to work toward redemption—however long it might take.

I passed my time in confinement by working, exercising, and reading. I quickly realized that if I spent my days sitting in my cell watching television, the hours would crawl by unbearably slow—and worse, I would become even more vulnerable to drowning in my own guilt. The isolation from friends and family only intensified the weight of it all, making each day feel heavier than the last.

Most nights, when I lay down to sleep, rest never came easily. My mind was always racing—thinking about how Kristen was getting along without me and replaying the nightmare of what happened to SPC Tackett back in Iraq. I couldn't stop picturing his family and the pain they must be enduring.

Eventually, the insomnia became so unrelenting that I had to submit a DA Form 510 to request an appointment with the Assistant Social Worker, Mrs. Wellman, hoping I could be prescribed something to help me sleep.

"How are you doing, Mrs. Wellman?" I asked as she motioned me into her office.

Mrs. Wellman was a short, pudgy, blond-haired woman in her early thirties whose warm demeanor seemed out of place in the harsh environment of the facility.

"I'm doing fine, Inmate Davis," she said with a polite smile. "How are you holding up back there?"

"I'm doing okay, I guess, Ma'am," I replied, though my voice lacked conviction. "But I'm having a lot of trouble sleeping."

"Why do you think that is?" she asked gently.

I took a deep breath. "Ma'am, I keep thinking about… and dreaming about… the way SPC Tackett died."

"What images are you referring to?" she asked, her expression shifting to one of concern.

"After I shot him in the head and he fell to the ground," I began slowly, my throat tightening as the memories surfaced, "I remember straddling him, placing my hands on his head to try and stop the bleeding. I keep seeing myself pull my hands away and look down… and his blood was all over them. That's how I've come to see it, Mrs. Wellman—his blood is on my hands."

She nodded, listening intently, her eyes soft with compassion but heavy with the weight of my words.

I also kept myself busy during the day. I had been handpicked by the Chief of the Prisoner Services Branch to work in the facility supply room after an incident involving the former supply worker.

One morning, I was called into MSG Lamoth's office.

"Inmate Davis, I think I have a job opportunity for you," he said, his tone serious but not unkind. "There have been some questionable practices happening in the supply room, and I'm looking for someone I can trust."

He leaned forward slightly, his eyes locking with mine. "Can I trust you to do the right thing and follow orders?"

"Yes, Cadre," I replied firmly.

"Good," he said with a nod. "On Monday morning at Work Call, report to Mr. Maragh in the facility supply room."

As instructed, the following Monday after the Work Call, I reported to the Facility Supply Room and met with Mr. Maragh and Sgt. Rogers, the Supply Sergeant.

They gave me a quick orientation on the expectations for my new assignment. As the inmate assigned to work in the supply room, I was responsible for a variety of functions essential to the smooth operation of the Regional Correctional Facility.

Each day began with inventory checks. I assisted in logging and organizing supplies—everything from uniforms and linens to cleaning materials and office equipment. It was my responsibility to ensure that incoming shipments were received properly, stored in their designated areas, and recorded accurately in the supply logs. I issued supplies to staff and inmates as needed, maintaining accountability for every item that passed through my hands.

The work was monotonous but strangely grounding. It gave me a sense of routine, a purpose to anchor me amidst the chaos of confinement. Sgt. Rogers made it clear from the start that the supply room had recently experienced issues with misuse and mismanagement, which was why I had been personally selected for this role.

"I need someone dependable back here," Sgt. Rogers told me firmly. "Someone I don't have to second-guess. Can I count on you, Davis?"

"Yes, Cadre," I replied without hesitation.

In that moment, I understood that this assignment wasn't just about folding sheets or stacking boxes—it was an opportunity to prove, even in a small way, that I could still be trusted. It was a chance to rebuild some measure of discipline and integrity within myself, qualities I felt I had lost the day my negligence cost SPC Tackett his life.

Though my days in the supply room were long, they provided me with quiet moments to reflect and to start the slow process of piecing together the fragments of the man I once was.

12

CONVERSATIONS WITH GOD

Psalm 27:8:

"My heart has heard you say, 'Come and talk with me. ' And my heart responds, 'Lord, I am coming. '

If any life-changing situation can bring a person closer to God and to His divine purpose and spoken word, this was certainly one of them.

Atheists and agnostics often argue that belief in God exists primarily to bring comfort and stability to humanity— a way to explain the unexplainable by clinging to the notion of the unexplainable. But I've come to realize that just because something is beyond human understanding does not mean that God does not exist.

Even in the depths of confinement, I could feel His presence. I witnessed how He moved in my life, even when I was surrounded by concrete walls, steel bars, and the weight of my own guilt.

My spiritual journey and my relationship with God became more profound during my time of confinement at Fort Knox. With my family miles away from me, both physically and emotionally, God became my only refuge. He was the one constant I could turn to, the only presence that felt unshaken by my failure and shame.

Leading me along this spiritual path was the facility chaplain, Chaplain Jones. He became a guiding light during some of my darkest days, offering words of encouragement, prayers, and a gentle reminder that even in my brokenness, God had not abandoned me.

I realized that if there was one thing I needed—and wanted—more than anything else, it was divine intervention. I knew that only God could give me the strength to endure this season of isolation and remorse.

I began attending church almost every Sunday while confined. It was the only place in the entire facility where I felt truly free—free from the heavy labels of inmate and felon, free from the suffocating weight of my own guilt. In that small chapel, surrounded by other broken men seeking their own redemption, I felt less like a recluse and more like a child returning home to his Father.

In search of answers, I went to church and attended spiritual counseling sessions with the Chaplain. At first, I was mad and furious with myself for not paying attention to detail and straying from my duties. But then, as my thoughts and my mind wandered in every direction, I began to direct some of that anger towards God and became angry with Him for allowing such a devastating event to transpire, and for allowing me to become negligent in my actions and of my duties in following Army regulations concerning my weapon. Finally, I became angry at God for allowing SPC Tackett to die and cause unknown hurt and devastation to an untold number of people. Where was the divine intervention at?

I had asked that question more times than I could count. If God was truly with me—if His hand was on my life, then why didn't He stop me? Why didn't He whisper louder, or slow my actions, or put something in the way to prevent what happened? I believed in providence. I believed in purpose. But in that moment, and in the long nights that followed, I struggled to believe that any of this was part of a plan.

I wanted the blame to be elsewhere, not entirely on me. And for a time, God seemed like the logical choice. If He were sovereign, if nothing happened outside His will, then surely this tragedy couldn't rest solely on my shoulders. That line of thinking gave me a brief sense of distance from the crushing weight of guilt. But it didn't last. Because deep down, I knew the truth.

And yet, in the depth of my conscience—in that quiet, aching space where no excuses could survive—God was the only place I could turn. Not to assign blame anymore, but to plead for mercy. I wasn't looking for a loophole. I was looking for a lifeline. I needed something—Someone—strong enough to face the full truth of what I had done, and still not turn away from me. And in that sacred desperation, I began to find Him.

I frequently sought out the spiritual counsel of Chaplain Jones while confined. Chaplin Jones was a young Black man who seemed to have a good heart and knew how to relate to me. He told me some of his personal history, and I realized that his upbringing was really no different than mine. As a Chaplain, he was different from the rest of the Cadre. He was able to extend his hand and have physical contact with the inmates and console them when needed.

I first met him when I was in-processing at the facility. I walked into the facility chapel and into his office. "How are you doing, Sir? I am Inmate Davis, and I am here to in-process, Sir." "Hello, Mr. Davis," the young-looking African American Chaplain replied. It was shocking to me that he did not call me "Inmate." He said Mr. It was a welcome recourse from the cadre addressing me as "Hey inmate," or "Inmate Davis report to the bench," which was where inmates had to report to before they were escorted by a cadre member to their destination in the facility, whether it be social work, the supply room, visitation, or to the chapel. "Come in and sit down, Mr. Davis," he said as he motioned me into his office. I stepped into his office and sat down in a chair adjacent to his desk. "Well, how are you doing today? How are you adjusting to the facility, Mr. Davis?" He asked. And I would answer with the normal response. "I am fine, Sir, I replied." Even though I knew that I was not okay. I was not getting any sleep because of the horrible thoughts and dreams I was having. I hated the fact that I was confined and away from my family, and I had thoughts of suicide constantly rolling around in my head at almost every waking moment. But this was how I responded to him and to others when they would ask me how I was doing and how I was adjusting to life in military confinement.

"And I am adjusting the best way I can adjust to life in here. I am just taking it day by day."

"Well, then that is good to hear. Have you been in contact with your family since you've been here?" he continued.

"Yes, Sir, I have been in contact with both my wife in Atlanta and my parents in New Jersey."

"Very good, very good," he stated.

"Do you have any children, Mr. Davis, and how is your family adjusting to your confinement?"

"I am married with no children," I stated.

There was a group of inmates who went to church for

a multitude of reasons. Whether it was to just get out of the cell and away from the cadre and/or other inmates, or just to go to church to seek the word and spiritual fulfillment, it was a welcome relief from the perils of confinement. I went to church for forgiveness. In the pew at the facility chapel, I sat next to criminals and sex offenders who all committed intentional criminal acts, some stupid and some I considered extremely heinous. Things I would never ever do. But I would sometimes think about what brought me to the Fort Knox Regional Correctional facility. I was here, and I would conclude that even though I was surrounded by former soldiers who served their country, they made bad decisions. These were people who, if I was not confined, I would never be in the presence of or even want to know. I figured, "Who was I," to judge them for what they did to get here. God, I believe, judges us all for what we do in this life, and I thought that maybe He should judge me the hardest. Placed in a position of responsibility, in which people's lives were in my hands, I failed to do for myself what I was enforcing and supervising in my soldiers. Especially with a weapon that was meant to take life.

13

THOUGHTS PERCENTAGES OF BLAME

THE BLAME GAME

ESV James 1:12:

Blessed is the man who remains steadfast under trial, for when he has stood the test, he will receive the crown of life, which God has promised to those who love him.

Throughout history, humanity has demonstrated an age-old tendency to deflect responsibility, a pattern that has played out in countless narratives—and the pages of the Bible are no exception. Scripture repeatedly warns against the dangers of blame and calls believers to embrace personal accountability for their actions.

The very first instance of blame-shifting occurs in the Garden of Eden. After eating the forbidden fruit, Adam deflects responsibility by pointing to Eve, saying, *"The woman whom You gave to be with me, she gave me of the tree, and I ate"* (Genesis 3:12, NKJV). Eve, in turn, blames the serpent: *"The serpent deceived me, and I ate"* (Genesis 3:13, NKJV).

These verses illustrate how blame can be used as a

shield to avoid confronting the consequences of one's own choices. Rather than acknowledging their disobedience before God, Adam and Eve sought to redirect responsibility elsewhere. But in doing so, they failed to recognize the transformative power that comes with confession, repentance, and accepting the weight of one's own actions.

The lesson remains timeless: blame may offer temporary relief from guilt, but it never leads to true healing or restoration. Only when we take ownership of our failings can we open the door to God's grace and redemption.

Many nights, my thoughts and dreams would lead me to wrestle with a host of questions about the Iraq War—questions about why we invaded Iraq in the first place and why Saddam Hussein was ousted from power.

I realize that what I am about to say may be controversial, especially to some within the military and government. But after everything I have experienced—after enduring this tragedy and living with its consequences, my support for the war, which had already begun to waver when I was commissioned as an officer in 2003, has now completely eroded.

In the wake of SPC Tackett's death, I find myself fiercely opposed to it.

As I write these words today, over 4,322 United

States military men and women have been killed in action—or by some non-combat-related incident—and more than 31,000 have been wounded in one way or another.

SPC Joseph Tackett was the one-thousand-eight-hundred and second soldier to die in that conflict. But his death was not the result of enemy fire, an IED, or the chaos of combat. His life was cut short because of *my* gross negligence, my failure to uphold the most basic standards of leadership and attention to detail.

As a commissioned officer in the United States Army, I had sworn an oath to lead American soldiers to victory in whatever mission we faced, in both peacetime and war. I was entrusted to protect them, to set the example, and to uphold the sacred trust placed in those who wear the rank.

Even now, years later, I am still learning to live with the reality that I failed—failed my country, failed my fellow Americans, and most painfully, failed the soldiers I was entrusted to lead in Iraq.

SPC Tackett did not make the ultimate sacrifice for his country. He was not taken by enemy hands in a distant land. Instead, he was shot to death by one of his own lieutenants.

That, I believe to this day, is a shame before God.

As I slowly began to come to terms with what I had

done—and with how foolish and careless I had been on the night I killed SPC Tackett—I found myself caught in a relentless tug-of-war within my own soul.

Part of me still longed for death, convinced that suicide was the only way to atone for the life I had taken. I believed that ending my own existence might serve as some form of retribution, a final apology carved out in blood for SPC Tackett and his family.

But another part of me, a larger and more stubborn part, wanted to live—not for my own sake, but to ensure that SPC Tackett's death would not be in vain. That part of me clung to the hope that somehow, through my words, my actions, and my choices moving forward, I might redeem myself, however imperfectly, for what had transpired that night.

I was torn between two competing truths: the crushing weight of guilt that whispered I deserved no future, and the quiet but insistent call to live a life of purpose, shaped by redemption and a relentless desire to honor the man whose life I had taken.

Moreover, over the last few years, I have become a prisoner—not only of my nightmares, but of my own mind. As I grieved more deeply for SPC Tackett, my thoughts became consumed with the war itself. Night after night, I watched the conflict drag on across television screens and read countless newspaper articles detailing the progress—or lack thereof—of Operation Iraqi Freedom. I listened as the Bush Administration's reasoning and rationale for waging war in Iraq were challenged by the media, political leaders, and others, even as the situation on the ground continued to unravel.

To date, I have lost seven friends to this conflict—seven lives cut short in a war that increasingly felt senseless. I thank God they brought in General David A. Petraeus in 2007 to implement the surge and attempt to stabilize the chaos. But even with that, I couldn't ignore the gnawing truth rising within me.

I began to shift some of the blame toward the leaders who had initiated what many now called a "war of choice." My anger grew as I reflected on the decisions that led us into Iraq, and even now, I remain infuriated over our country's involvement in this conflict—a war that cost so much and gave so little in return.

Now I am going to make this emphatically clear! I was negligent in my duties as far as handling my M16A4 weapon. The blame for SPC Tackett's death is rightfully placed squarely on my shoulders and is wretchedly embedded in my consciousness for the rest of my life. I made an inexcusable error in judgment and two major mistakes:

1. Every soldier must know the status of his or her weapon at all times—whether it is loaded or unloaded, and whether the selector switch is set to safe or one of the two firing positions: semi-automatic or burst. This is a fundamental rule, drilled into every soldier repeatedly from the very first day of weapons training. It is not merely a guideline; it is a standard of discipline and situational awareness that can mean the difference between life and death.

2. **NEVER point a weapon—loaded or unloaded—at anything you do not intend to destroy.** The reality is simple and brutal: what comes out of the business end of that weapon will kill. A weapon does not differentiate between enemy or friend, combatant, or innocent. As an officer, it was not only my responsibility to understand this truth but also my sacred duty to enforce it with absolute seriousness.

In this, I failed, and SPC Tackett's death is the result of that failure. However, while in confinement and even now as I write this memoir of sorts, when I lie in bed at night or when I watch the news and see new evidence emerge about the falsity of the rationale that led to the American invasion of Iraq. I began to think that this would not have ever happened if we had never invaded Iraq.

I believe that Saddam Hussein was a tyrant and a murderer, and what he did to his people during his reign as the President of Iraq was unfathomable to the average human being. It is evident that he and his minions did what they wanted and broke nearly every international law concerning human rights and United Nations Resolutions concerning biological and chemical warfare, as he inflicted unnecessary harm on his own people. The absence of evidence that no weapons of mass destruction existed, which President Bush, Secretary of Defense Donald Rumsfeld, and Vice President Richard Cheney so emphatically stated, has yet to turn up, and the connection between Saddam Hussein and **Al-Qaeda** is increasingly becoming an interesting fallacy.

I always think about what-ifs.

What if I had cleared my weapon and properly followed the clearing procedures that I was trained to both perform and enforce?

What if I had adjusted my tactical sling so that the muzzle of my weapon would always point to the ground so I would not have to grab the handle and attempt to force it down?

What if the United States had never invaded Iraq, and how would life have been different?

Would SPC Tackett be alive and well today? My heart tells me yes. I'd like to believe he'd be thriving, maybe back home with his wife, starting a family, building a future he'd earned through service and sacrifice. I imagine him laughing with friends, pursuing his goals, living the life that was so unfairly cut short. That thought brings both comfort and pain, because I'll never truly know.

I honestly believe that if not for my failure, 1LT Kevin Smith would never have had to take my place, and he would still be alive today. He died on December 8, 2005, just weeks before the 3rd Infantry Division was scheduled to redeploy home, its mission all but complete. That burden, too, is one I carry.

I never had the chance to know First Lieutenant Kevin Smith well. I understand that he was a solid young officer, the kind of officer any unit would be lucky to have. What I do know is this: he died serving in a position I once held. He stepped into a role I had been removed from. And for that reason alone, I carry his name in my heart with the same heaviness I carry SPC Joseph Tackett's.

Some may call it a coincidence; others call it fate. But to me, it felt like a direct consequence, a chain reaction set in motion by my failure. When I was removed from command and sent to confinement, someone had to fill the void. That someone was Kevin. And three weeks before the 3rd Infantry Division was scheduled to redeploy home, he was killed in Iraq.

I can't say for certain what would've happened had I remained in the theater. Perhaps Kevin would still be alive. Perhaps I would have taken the mission that cost him his life. I'll never know. But that's the burden of leadership, our actions, our failures, our lapses in judgment don't just affect us in the moment. They echo. They ripple outward. And sometimes, they find their way into the lives of others, we never meant to harm.

In the military, we talk often about "the chain of command," but rarely do we talk about the chain of consequence. That invisible thread that links us to one another, officers to enlisted, commanders to subordinates, predecessors to replacements. When I failed in my duty, when I neglected to enforce what I had been trained to do, it didn't just affect one night or one soldier. It set things in motion I never anticipated. My absence created a vacancy, and that vacancy required another life to step forward.

I believe in God's providence, and I don't pretend to know His reasons. But I know what I feel. I feel responsible. Not by law, but by connection. Kevin Smith and I were bound not just by rank or role, but by the terrible handoff that occurred when I left in shame and he entered it in honor.

The death of SPC Tackett broke something inside of me. The death of 1LT Smith reminded me that leadership is not something you stop owning just because you're no longer in command. We don't get to walk away clean. We are connected—to the lives we lead, to those who replace us, and to those we never get the chance to thank or say goodbye to.

Lord knows, I would have gladly taken his place. To this day, I wish it had been me instead of him.

And because of that, 1 feel the weight of responsibility for Lt. Kevin Smith's death as well.

In my counseling sessions with the facility social worker, we often spoke about fault, blame, and the endless cycle of "what ifs" that consumed my thoughts. We also discussed the delicate question of timing and the legal framework for approaching the Tackett family—when and how I might let them see the face of the man who had once been entrusted as one of their son's superiors, yet failed to uphold his most basic duty in a seemingly simple task.

That failure had resulted in the taking of their son's life.

I didn't know how to process that in my mind. The weight of it felt heavier than any physical confinement. It was as though my body was imprisoned in walls of concrete and steel, but my soul was locked in a prison far more suffocating—one built of guilt, regret, and shame.

"Sir, I'm living with something that feels almost impossible to bear," I said, my voice low and heavy with emotion. "Since we last spoke, I've been haunted by relentless nightmares and plagued by recurring thoughts of ending my own life. It's overwhelming, I'm caught in a nightmare I can't wake up from, no matter how hard I try.

Never in my life did I think I'd find myself in this position. I've considered whether taking my own life would be the only remedy, a way to release myself from all the pain I've endured these past few months.

But at the Fort Knox Regional Correctional Facility, even that felt impossible. Privacy didn't exist—not for a moment. No matter where I turned, I was surrounded by inmates or under the watchful eyes of the guards.

Still, the thoughts came often and uninvited. I even tried to imagine how I might do it. Hanging myself in the inmate shower crossed my mind more than once. But even there, it was impractical. The flimsy shower rods couldn't have supported my weight, and the thin curtains separating the stalls offered no protection from being seen. Guards were constantly patrolling the track outside; even if I attempted it, the chances of succeeding were slim.

As I sat across from Mr. Bowen during our sessions, I found myself voicing these thoughts aloud for the first time. He sat and listened with a steady, almost disarming calmness, occasionally offering quiet words of advice.

After a few months in confinement, I began to realize something I had never utterly understood before: for me, confinement didn't just restrain my body, it waged war on my mind. It pressed in on my thoughts, distorted my sense of time, and magnified every emotion, especially guilt. The weight of it burrowed deep into my soul, chipping away at the man I thought I was and reshaping me into someone I barely recognized.

14

RECONNECTING, LIFE AFTER KNOX

Matthew 6:34:

"Therefore, do not worry about tomorrow, for tomorrow will worry about its own things. Sufficient for the day is its own trouble."

I was released from confinement from the Regional Correctional Facility at Fort Knox on the morning of 1 July 2007. The previous night, I slept better than I had slept for a long time. For the first time in a long time, I did not even use the prescribed sleep medication I was taking for me to sleep through the night. I was so happy to be going home that I pretty much rocked myself to sleep. God, I was so happy and elated to be getting out of confinement and to move on with my life and to be reunited with my wife. Which was both exciting and frightening to me, I loved my wife very much. I could not wait to be able to hug her and not let her go. No more Cadre telling me that time was up, no more wondering if Kristen was going to come and see me or not; it was a huge relief.

I was leaving with two other inmates, neither of

whom served nearly as much time as I did. One of them only served for three weeks. As I heard him say, "Man, I am so glad to be getting out of here, three weeks is a long time." "Give me a break," I thought to myself. I spent twenty-two months, six hundred and seventy-one days in confinement.

"Control, Guard Commander," SSG Regan stated over his radio as I and two now former inmates walked out of the doors of the Regional Correctional Facility at Fort Knox. "Set population to 133! Inmates Wilcox, Andrews, and Davis are released, effective 0759." I heard SSG Regan state calmly. A PDF Guard loaded us up into a white van, and we went down the street to the soldier holding facility awaiting pickup. As we drove down the street, I faintly saw my wife's car and thought to myself that I was free.

For the first week after I was released, I enjoyed my wife, or I should say that I tried to enjoy her. When we pulled up to our apartment in the Lyndon neighborhood of Louisville, Kentucky, I felt that I was restarting my life all over again. However, I also felt unsure and scared. I did not know what I was going to do, and I did not know how to go about doing anything. I knew that I would continue to have personal struggles, but I felt that I would be strong enough to confront what would come my way, it became much harder than I had anticipated.

As we began to approach Louisville, Kentucky, exiting the Gene Snyder Parkway and onto Interstate 264, I was happier than I had been in over two years. We pulled into an apartment complex called Post Oak Apartments, in which my wife had rented a one-bedroom loft, the second of two apartments she had rented soon after moving to Kentucky in order to be closer to me and to better support me. "This is nice," I said with enthusiasm as I entered our apartment for the first time. "I know, we have to get a bigger place now that you are home," she said as she leaned in to give me a kiss. She began to give me a tour of the apartment, as she showed me our pictures together and a collection of chess sets that she had brought for me.

After the tour, we went upstairs to the bedroom and laid down. I was able to really hold her and hug her without the Cadre at Fort Knox telling us not to display too much affection and not to touch. It was really good to be home and good to be in her arms. I thanked God that I was married to a woman and that she stayed with me faithfully and supported me during my time in confinement. I thought that even though I was responsible for the death of SPC Tackett. The Lord still showed favor to me enough not to take my wife out of my life. And I felt grateful for that blessing.

Almost immediately after coming home, I began

searching for a job in the Louisville area. Since Kristen needed the car most days, I had to rely on the local public transportation system to get around.

Now, I was a convicted felon with a dismissal from the United States Army hanging over me like a dark cloud. It wasn't just about finding a job—I was worried about finding one that paid enough to support us. Deep down, I knew I wouldn't be earning anywhere near what I had made as an Army officer, but I also knew I had to do something.

One evening, as we sat together, I turned to Kristen with a heavy sigh.

"Kristen," I said hesitantly, "how's the job market here in Kentucky? And... what kind of job do you think I can even get with this felony on my record? I mean... I'm going to have to explain what happened every time I apply for something. I guess I just have to get used to that."

She paused for a moment, her face calm but thoughtful.

"Well," she began, her voice steady, "I think the first thing you need to do is start getting the local newspaper every day and check the employment section. Then start applying for jobs and going on interviews."

"There is work out here, Will. No matter what job you get, it's true you're not going to make what you were used to making as an officer in the Army—but that's okay."

She reached over and gently touched my hand. "I suggest you start with a part-time job, just to get your foot in the door. Start making some money, and we'll go from there. We'll figure it out together.

Her words were calm, reassuring—but I could still feel the weight of reality pressing down on me.

So that's exactly what I did. Each morning, I got up and went out to submit job applications at every retailer I could find, requesting interviews with various department stores and businesses—mostly for positions in sales and customer service.

Kristen would get me a newspaper, and I would circle the jobs that I was interested in and then call them. Kristen also helped me get my resume together and helped me prepare for job interviews. I did not care what job I got. Whatever I got, I looked at it as a step forward. I had to make a living doing something, even if it meant working a job that I did not particularly want to do or even cared for. I figured that working would at least take my mind off the thoughts that invaded my mind almost daily. It would at least keep me busy from the thoughts of suicide that were beginning to really creep into my psyche.

I thought that I would be okay once I was released

from confinement and moving on with my life. But as I tasted freedom once again, feelings and thoughts that I did not expect began to take hold of me. For one, I was back home with my wife. I was broken and unsure of everything. But I felt that I did not deserve that or even to be alive. In moments that should have been filled with love, I felt haunted. I imagined Tackett—alive, loved, and laughing with his wife—and I would be overcome with the weight of what I had taken. Kristen would look at me sometimes and would ask me what was wrong, and I would lie to her and say, "Nothing, nothing." But a lot would be wrong with me. My consciousness would get the best of me sometimes, and I began to think heavily about what Tackett's life would be like if he were alive. How his parents and his wife would have welcomed him home after he stepped onto the tarmac at Hunter's Army Airfield after a one-year tour in Iraq. Welcoming him home with open arms, huge smiles, and tears of joy. And even though I was in confinement, I still made it back home to be in Kristen's arms a couple of years removed, and to be happy with my family. It just did not seem right to me. I went to church, and I would listen to the sermons that were preached and filled with words after each sermon. But I was still struggling with life. Most of all I was struggling with my existence.

PART IV:
REDEMPTION
AND
RESPONSIBILITY

15

THE COST OF MY CARELESSNESS

"One reckless moment can ruin your entire life, always remember that, Willie."

Those were the words my mother repeated to me often throughout my childhood. At the time, I didn't give them the reverence they deserved. They were warnings from a loving mother, words spoken to guide a young boy who didn't yet understand the consequences of his actions. But I now know that nothing could have prepared me for the devastating weight of those words until the moment my carelessness caused the death of Sergeant Joseph Tackett. That one act, one tragic, preventable act, forever transformed the course of my life.

After my release from confinement, I struggled to rebuild what remained of my future. The road to redemption, if one even existed for me, began with the humbling task of finding work. Eventually, I secured a part-time position at an office supply store that served small and mid-sized businesses. It wasn't glamorous. It wasn't what I had dreamed of as a young ROTC cadet or commissioned officer. But it was something, and I was grateful. I learned to find small victories in simple successes: connecting with customers, upselling warranty plans, and showing up every day. For someone freshly released from military confinement, even that was no small feat.

Kristen and I quickly realized that our small, shared loft no longer suited the new shape of our lives. I longed for something modest, a quiet apartment, something easy to manage. Kristen, ever hopeful and optimistic, dreamed of a home. A house, she believed, would symbolize a fresh start. Though I worried about my limited income and emotional fragility, I wanted to support her dreams. So, despite my fears, I agreed.

By the end of September, we purchased a shotgun-style home near downtown Louisville. The house was beautiful in its simplicity, a lovingly restored structure full of charm. On paper, my life looked like it was coming together: a job, a home, a wife who had stood by me during the darkest period of my life. But inside, I was unraveling.

Almost immediately, our marriage began to show signs of strain. I was emotionally distant, disconnected from joy, and crippled by guilt. Night after night, I woke up drenched in sweat, haunted by the image of Joseph Tackett. I'd sit in silence at social events, unable to laugh or enjoy myself. To me, each smile felt like a betrayal of his memory.

How could I enjoy life when I had taken another man's?

Kristen noticed. She would ask, "Why don't you talk to anyone?" But I never had the words. I couldn't articulate the depth of my self-loathing, so I simply said, "I don't know." I tried to shield her from my pain, but in doing so, I shut her out completely.

The weight of what I had done grew heavier with each passing day. I thought not only of SPC Tackett, but of his family, his mother, especially. I still remember her picture on the front page of the Louisville Courier Journal, tears streaming down her face. I imagined losing my own son that way. The grief, the confusion, the rage, it was unbearable even to think about. But she had to live it.

This wasn't an accident born of recklessness like driving drunk. I was sober. I was trained. I knew the rules. And yet I still failed. I played with the safety selector switch. I neglected muzzle awareness. I walked into a safe zone without clearing my weapon. All those mistakes were mine alone.

Before I was released, I prayed that Kristen and I could make it through. I prayed our marriage would survive. But I knew deep down that things had changed. The man she picked up from Fort Knox on July 1, 2007, was not the same man she had married.

I tried to be honest. I offered her my Record of Trial, hoping she would understand. I thought it would bring us closer, that she would see the pain I carried and stand with me in it. But when she finally read it, her reaction wasn't what I expected.

I walked into the living room one afternoon and saw her sitting on the couch, reading through the thick blue packet. "How could you flag your soldiers with your weapon?" she asked, her voice laced with disbelief.

I tried to explain, but she looked at me with disgust. It was the kind of look that says everything words cannot. It was the look of someone who no longer recognized the person sitting across from them.

From that moment on, I stopped confiding in her. I had always thought she was my safe space, my refuge. But I no longer felt safe sharing my thoughts. The shame was too great.

Years earlier, I once told her, "I could be a general someday." I was proud of my career in the United States Army. Her response stunned me. "Willie," she said flatly, "you'll never be a general."

She was right. I wasn't general material. I was a disgrace. A failure. And now, I have lost the respect of my wife too. The one person I thought might still believe in me.

We sought counseling at her urging. Our therapist, Rick, was a Gulf War veteran who understood my world in a way few others could. He helped. But even with his support, I couldn't bring myself to open up to Kristen again.

I began to retreat further into myself, questioning whether life was still worth living. I had taken so much and lost even more. I lost the respect of my soldiers, my peers, and now my wife. What else was there?

I started thinking about suicide more seriously. The pain was unrelenting. I was exhausted. My worst fear that my marriage would not survive was coming true. Kristen didn't call me stupid, but her disappointment was palpable. And in my heart, I felt stupid. I felt like a failure. A burden.

She deserved better, a man without baggage. A man who hadn't brought shame to her doorstep. I wanted to let her go. Maybe that's why I shut down. Maybe that's why I stopped trying.

Even when we tried to talk, her words were cutting. "How could you be so careless?" "What was wrong with you?" "Why did you do it?" I didn't have an answer. I still don't. I just know it hurts.

Even now, years later, I carry it with me. Every single day. Some days are lighter than others. But the guilt is always there. It sits with me in silence. It lies beside me at night. It shapes how I see the world and how I see myself.

I don't know if I'll ever be free from it. But I'm still here. Still trying. Still breathing. And maybe, just maybe, that's enough.

16

THOUGHTS: AN OFFICER'S BETRAYAL OF OATH

Romans 12:2:

"Do not be conformed to the pattern of this world but be transformed by the renewing of your mind. Then you will be able to test and approve what God's will is, his good, pleasing, and perfect will."

On May 17, 2003, I took the following oath:

"I, Willie Davis Jr., do solemnly swear that I will support and defend the Constitution of the United States against all enemies, foreign and domestic; that I will bear true faith and allegiance to the same…"

I said those words with conviction. I meant them with every fiber of who I was. I was proud, eager, and determined to lead—to serve with integrity and courage. That oath wasn't just a formality. It was a commitment I believed in down to my bones.

But later, confined at Fort Knox, that same oath echoed through my mind like a hollow bell. I questioned what it meant anymore—what I meant anymore. Lying awake at night on a stiff military cot, I often stared at the ceiling and asked the same question over and over: How did

I let this happen?

The Bible was always nearby. At times, it was a source of strength. At other times, it brought grief so sharp I couldn't breathe. One night, I found myself reading about Judas Iscariot. I had heard the story a hundred times in Sunday school, but that night it read differently—personally.

Judas, the man who betrayed Jesus with a kiss. Who, for thirty pieces of silver, handed over the innocent. Some scriptures say it was greed. Others say Satan entered him. But what stayed with me was this: Judas made a decision he couldn't undo. He tried to return the silver, but it was too late. He couldn't bear the weight of what he had done. And he took his own life.

I read that passage, and for the first time, I didn't just see a villain—I saw myself.

Because what happened in Iraq was not a tactical order. It wasn't a mission gone wrong. It was a grossly negligent mistake—my mistake. And because of that, a good man died. SPC Tackett lost his life, and I was the officer responsible for the safety and welfare of every soldier under my care. I failed him. And that failure still haunts me.

Joseph wasn't a name on a casualty report. He was a 22-year-old soldier from Whitehouse, Kentucky. A son. A husband. A man full of life and plans. His mother, Kathy

Tackett, said he had dreamed of joining the Army since high school. He wanted to see the world—and he did. He wrote home about the mosques he visited, the things he was learning, and the people he met. He told his mom, "There aren't many people who've ever done this, Mom."

He was stationed in the Green Zone, escorting dignitaries, offering letters of encouragement to schoolchildren back home—his "pen pals." He loved skateboarding. He shared American rock music with Iraqis. He was curious. Kind. Full of promise.

His mother said something that has never left me:

"I can't imagine the person that he would've become, if he had more years."

And neither can I.

What happened to him was avoidable. That truth is an unbearable burden I will carry for the rest of my life. There were days in that cell where I thought the only way to pay for it was to follow the same path Judas did—to end the story myself. The enemy whispered to me in those moments, telling me I was too far gone, that no grace could cover a mistake like mine.

But grace—God's grace—isn't logical. It isn't earned. It doesn't excuse the consequences, but it enters into them with compassion. And somehow, it found me there.

Unlike Judas, I didn't take the ultimate step into the darkness. I clung to that thin, trembling thread of mercy. Because I began to realize something essential: SPC Tackett's death must not be met with another death. It must be met with truth, with humility, and with a lifelong commitment to honor the man we lost.

I live today not because I deserve to—but because I have a responsibility to remember, to speak, and to serve others so that no leader makes the same mistake I did. Joseph's life was sacred. It matters. And maybe, just maybe, my story is one small way I try to carry his name forward.

God's grace didn't erase what happened. It didn't take away the weight of my failure or the pain I caused. But it gave me a reason to live differently—a reason to move forward with purpose.

And that reason began with one young man's life—full of promise, full of potential—silenced far too soon.

17
MAKING SENSE OF IT ALL

Isaiah 43:25 (ESV):

"I, I am he who blots out your transgressions for my own sake, and I will not remember your sins."

It is written that the Lord will blot out our transgressions, choosing not to remember our sins, for His sake, not ours. But I am not the Lord. I am a man. And I remember my sins. I remember them every single day. Each morning, I rise with a shadow over my soul, a heavy silence that no sound can drown. And at night, when the world quiets and there is no one left to distract me from my thoughts, I revisit those painful, brutal, and unrelenting memories. I have come to learn that when life is shattered by tragedy, when a soul is crushed under the weight of grief and remorse, a person stands at a crossroads with only two possible paths.

The first path is the path of surrender. It is the slow, agonizing descent into hopelessness, a road paved with guilt, shame, and self-disgust. It's the conviction that one's life no longer holds value or purpose. It's waking up and wishing you hadn't. It's looking in the mirror and seeing a stranger whose presence you can no longer tolerate. I walked that path for the better part of three years. And there were many nights, more than I care to admit, when I stood at the edge of existence, staring into the abyss, contemplating if stepping in front of a moving bus or under the wheels of a semi would bring the release I thought I deserved.

But then there is the second path. The harder path. *The path of faith.* The path of slow, painful adaptation. It is the decision not to give up, not to allow your grief to destroy you. For me, it was the hand of God that led me away from that darkness. It was the quiet but persistent presence of faith that offered me another chance, not a chance to forget, but a chance to live despite remembering. I passionately believe that the Lord placed certain people in my life precisely when I needed them most, people who, knowingly or not, pulled me back from the brink. Their compassion, their patience, their refusal to let me give in to despair, these were divine interventions in human form.

I once stood in a military courtroom in Baghdad, facing judgment for the unthinkable: the negligent taking of a fellow soldier's life. The judge reminded me that even accidental death is still the taking of life, and that all life is sacred under the law and before God. No one gets a pass for taking a life, not even when it wasn't meant. That day, something inside of me solidified: I had to find a way to live with myself. I had to find purpose in the wreckage of my life. Because that bullet, fired from my weapon, a weapon I believed was unloaded, took more than a life. It shattered a universe.

You don't easily forget seeing a man's eyes after he's been shot in the head, especially when it's by your own hand. I see them still, those eyes filled with confusion, disbelief, and fading life, asking me silently, "How could this happen?" I failed him. I failed his family. I failed my platoon, my uniform, my country. And no rank or ribbon or sentence can ever change that truth.

I live now with that failure like a constant companion. Some nights, I wake drenched in sweat, my breath short, my heart racing. But my faith steadies me. It grounds me. It reminds me that while I cannot undo what happened, I can shape what happens next. I can strive to make the remainder of my life a tribute to him, a pursuit of justice, of humility, of service. Because that is what SPC Tackett deserves.

St. Thomas Aquinas once wrote, "Justice is a certain rectitude of mind whereby a man does what he ought to do in circumstances confronting him." Those words live in me now. They remind me that justice, in this case, cannot be served through legal means alone. No prison sentence, no demotion, no apology can match the weight of a life lost. True justice must come from within. It must be pursued through action, through a life lived in dedication to something greater than oneself.

In the wake of this tragedy, I have found that the human soul, when wounded deeply, will reach out for meaning. We search for it in order to keep going. To move forward without understanding is to wander aimlessly through darkness. I believe we all need a reason, a compass, a guiding star. In my case, that reason is twofold: to honor SPC Joseph M. Tackett, and to find peace for myself.

Every day when I wake up, I remind myself that I am not only living for myself. I am living for him. His death must not be in vain. If I ended my life, if I silenced this voice inside me, then what would that accomplish? Would that be justice? Would that be redemption? No. It would be a surrender. And surrender is not an option.

It has taken the unyielding love of my family and friends, and something deeper than them all, the presence of God, to help me withstand this daily struggle. In the depths of my soul, there is a battlefield, and I fight there each day: to stay faithful, to keep believing, to find worth in my existence. I battle against self-doubt, against the shadows of the past that still try to convince me I am not worthy of forgiveness or hope.

But I do have a purpose. And that purpose is to serve.

I cannot undo the past, but I can live in such a way that uplifts others. It took four years after the incident for me to fully accept the weight of my guilt, and even longer to transform that guilt into something constructive. But now, with time, I understand. I must give back. I must serve humanity in whatever ways I can, large or small, visible or hidden. Because each act of service is a stone laid on the road to redemption.

Living with this reality is both a blessing and a curse. There are good days when I feel strong and clear-eyed. There are bad days when I can hardly stand. And then there are those days that fall somewhere in between. But I continue. I press on. Because I must. Because Tackett cannot.

Without my belief in God and the possibility of divine forgiveness, even for something as grievous as this, I would be utterly lost. I believe that the Lord still walks with me, even when I stumble. I believe that He has not forsaken me. And the people in my life, old friends and new, continue to show me what it means to be loved even in the aftermath of devastation. Their love, their loyalty, and their grace have helped me inch closer to forgiveness, not just from God, but from myself.

I now understand that I must live a life that is worthy of the man whose life was cut short. And the only way I know how to do that is to dedicate myself fully to the service of others. That is how I will honor him. That is how I will survive. That is how I will heal. If I can ease the suffering of another human being, no matter how small the gesture, I will have found purpose. I will have made sense of my life.

This is my path now. This is my road to salvation. The only one I have.

SPC Joseph M. Tackett once took an oath—to support and defend the Constitution of the United States, to serve as a soldier bound by honor and duty. I do fully grasp the weight of that oath now, because his sacrifice has forever been tied to my own failure. His life and service remind me that an oath is not just a pledge, but a sacred trust—one I mishandled, with consequences that can never be undone. Living with that truth has marked me deeply, shaping how I understand responsibility, sacrifice, and the burden of leadership. Honoring his oath now means carrying that weight with humility, faith, and the determination to live worthy of the cost.

18

MY GREATEST CASTIGATION

2 Corinthians 2:5–7:

Now, if anyone has caused pain, he has caused it not to me, but in some measure, not to put it too severely, to all of you. For such a one, this punishment by the majority is enough, so you should rather turn to forgive and comfort him, or he may be overwhelmed by excessive sorrow.

After all I have endured in the years following that harrowing day, after the trials, the institutional reprimands, the soul-searching, and the quiet implosions that have defined my life, my greatest castigation, my deepest punishment, was never the confinement at Fort Knox. Though I was confined for 22 months, living under the strict supervision of a military prison, surrounded by walls both physical and psychological, that sentence alone does not even begin to compare to the punishment I continue to carry inside me. In truth, the worst prison is not made of concrete or razor wire. It is constructed within the soul, in the form of guilt, memory, and relentless remorse.

During those 22 months, my mind wandered and spiraled into a hundred directions. I spent countless days lost in thought, searching for meaning, searching for understanding, searching for redemption. More often than not, I found myself circling around the same central, unanswerable question: What should have been a suitable punishment for my actions? What consequence could ever possibly fit the magnitude of negligently robbing another human being, SPC Tackett, of his life? Of stealing from his mother, her beloved son, from his wife, her partner and protector, from his siblings, their brother, and from his friends a man they respected and cherished. The list of the people affected by this tragedy is likely far longer than I can imagine, extending well beyond the few names I know. I will never know the full reach of the heartbreak my actions caused.

If the tables were turned, if I had been the one whose life was cut short that day, I know there would be people who mourned me too. Family, friends, brothers, and sisters in arms. People who would ask themselves why. People who would question how such a thing could happen. And yet that knowledge doesn't lessen the guilt I carry; it magnifies it. Whether SPC Tackett's death was an accident or not, it was entirely preventable. That's the truth I must carry. That's the truth I must never forget.

Let's take stock. Since June 23, 2005, I have not only been confined in a military prison, but I've also been forced to endure the psychological weight of being categorized alongside individuals who had committed deeply disturbing and intentionally harmful crimes, rapists, abusers, and child molesters. While I was not among them by nature or moral intent, I still had to exist beside them. I had to learn how to function in that environment, how to navigate that bleak terrain, how to find a way to survive and hold on to some sliver of humanity in the hope that one day, I could emerge from it all with the chance to rebuild, to try, however imperfectly, to redeem myself. I had to learn how to swallow the shame of being an officer who ended up in prison, and I had to carry it with humility in order to even begin to piece together the shattered remains of my life.

I tried to salvage my marriage, the one anchor I had before it all began to drift away. But that too, slipped through my fingers. My first wife, a woman who once believed in me, stood beside me and offered me her love and loyalty, but ultimately divorced me. And I can't blame her. I couldn't give her what she deserved. I was too broken. Too lost. Too consumed by guilt and self-loathing to be the partner she needed. There were times when I genuinely wanted to disappear from the world, to escape the crushing weight of it all by taking the easy way out. But I didn't. I stayed. I endured. And though our marriage did not survive, I remain grateful to her. I wish her nothing but love, peace, and happiness in all the days to come. I pray she is now embraced by the kind of joy and fulfillment I was unable to give her.

She once told me that our divorce wasn't entirely because of what happened on that day. Maybe she's right in a literal sense. Maybe there were other issues. But in this situation, how could I have been so sure? For the first few months that I was home after confinement, I was learning and trying to be a good husband for her. She deserved that after being away for the better part of three years. First, by living in two different residences separated by a three-and-a-half-hour drive between Fort Stewart, Georgia, and Atlanta; second, by enduring a deployment in Iraq, and third, my conviction by General Court-Martial of the grossly negligent death imposed by my own weapon and imprudent actions that resulted in SPC Tackett's death. She had a lot to put up with, more than what most people our age and in a young marriage have to put up with. So, I could not blame her at all.

But she could never fully understand that everything in my life after June 23rd, 2005, is a ripple effect of that one devastating moment. Every heartbreak, every failure, every sleepless night, every inch of self-doubt, every strained relationship, it all circles back to that day. That is the burden of living with the knowledge that your mistake cost someone else their life.

What I have reconciled with God, I have not fully reconciled within myself. Forgiveness, in the spiritual sense, is one thing. But self-forgiveness? That is a mountain I continue to climb, with no summit in sight.

So no, my greatest punishment was not delivered by the military justice system. It did not come in the form of legal sentencing or confinement. My greatest castigation is internal. It is deeply embedded in the very fabric of who I am now. It lies in the memories that refuse to fade, in the moments of quiet when the images return unbidden. It lives in the guilt that resurfaces every year as the date approaches, when the air seems heavier and the world feels dimmer, June 23rd, again.

I will never forget what happened. How could I? I was a commissioned officer in the United States Army, given the sacred responsibility to lead, protect, and guide the soldiers entrusted to me. And I failed in the most basic, most preventable way imaginable. I failed to clear my weapon. I failed to maintain muzzle awareness. I failed to adhere to the discipline and attention to detail that are the foundation of military life and leadership. I failed not only my men, but my chain of command, my service, and my country.

Some people try to comfort me by saying I am too

hard on myself. That I've paid my debt. That I should let go and move forward. But those people don't understand. They haven't seen what I've seen. They haven't felt what I feel. They didn't look into Sergeant Tackett's eyes in the seconds after the bullet struck him, those wide, shocked eyes, filled with confusion and pain, silently pleading, "How did this happen? Why did this happen?" They didn't cradle his bloodied head in their hands as his life slipped away, stolen not by an enemy, not by fate, but by a fellow soldier's moment of negligence.

I remember watching his eyes slowly roll upward, the light in them fading as his lids fluttered shut for the last time. That image haunts me. It doesn't just visit in the quiet of the night, it's etched into the back of my mind like a scar that refuses to fade. It returns in my dreams without warning—sometimes vivid, sometimes distorted—but always real enough to make me sit up gasping, sometimes wiping away tears. It echoes in my waking thoughts, playing like a broken reel at the edges of every moment of peace I try to find.

I'll be in the middle of a normal day, a conversation, a meal, a moment of laughter—and suddenly, there it is again: his face, his final breath, the slow dimming of the light in his eyes. It colors everything I do. Every decision. Every relationship. Every attempt to move forward. Because no matter how many days pass, I carry the memory of that life ending in front of me. A life that was never supposed to end that way. A life I was entrusted to protect.

That is my punishment. That is my penance. That is the castigation I carry, not just for a year, or a decade, but for the rest of my life. *Castigation*—a word that hardly scratches the surface of what I endure. This was no mere reprimand or rebuke handed down by a superior officer. It was an unrelenting, soul-deep reckoning that echoed louder in the silence than any sentence a courtroom could pronounce. It came in the still hours of the night, in the empty stares of my soldiers, in the haunting memory of a name I could not forget—SPC Joseph M. Tackett.

What I experienced was beyond discipline. It was spiritual scourging. A kind of internal crucifixion where the accuser, the judge, and the executioner were all me. I replayed every second leading up to that moment—every decision, every oversight, every instinct I silenced when I should have spoken. I questioned everything: my calling, my competence, my worthiness to even draw breath.

But buried beneath that grief—beneath the despair that almost claimed my life—was the distant whisper of a promise. That God could still do something new. That the story wasn't over. That even in this desolate valley, there could still be grace.

19
GOD'S GRACE

Ephesians 2:8–9

For by grace, you have been saved through faith. And this is not your own doing; it is the gift of God, not a result of works, so that no one may boast. For we are his workmanship, created in Christ Jesus for good works, which God prepared beforehand, that we should walk in them.

"What exactly is God's grace?" I would often think to myself. To me, the grace of God is His undeserved favor—His quiet, relentless kindness that shows up even when we least expect it, and especially when we don't deserve it. Grace is when He covers us with mercy instead of judgment. It's when He reaches down into the wreckage of our lives and chooses to restore instead of discard.

I've come to believe that grace isn't just about forgiveness, it's about presence. God's grace is in the beauty of the world we so often overlook, in the moments when accidents are narrowly avoided, in the right person or word arriving at just the right time. It's in the whisper of a conscience reminding us of who we are, and who we're still called to be.

These are what I've come to understand as common graces—gifts freely given to all of us, simply because God so loved the world. But I also know a deeper grace: the kind that walks with the broken, that refuses to leave you in your shame, and that patiently builds a new story out of the ashes of the old.

As I struggled to piece together a life I thought was over, I never imagined that God would still grant me grace. A couple of years after my divorce, I had quietly accepted that marriage was behind me—that I didn't deserve another chance. I believed no woman should love me after what I had done. I had already lost one woman to this tragedy, and I couldn't bear the indignity of bringing more shame upon anyone else. The weight of guilt had not only scarred me—it had closed me off.

I was reluctant to get involved with anyone. I had some "relationships" after my divorce, but none felt real, deserved, or sustainable. My heart wasn't in them. I didn't believe love could grow in a place so wrecked. But grace has a way of showing up when you least expect it. They say the Lord works in mysterious ways—and this was one of them.

After my divorce, I spent a few years trying to find myself. I was working at a home improvement store in Louisville, Kentucky, when I met one of the sweetest, most God-fearing people I've ever known. Her name was Priscilla. She was beautiful—inside and out. But it wasn't just her appearance that drew me in—it was her spirit. I felt pulled toward her like a magnet. Her presence was unlike anything I had ever experienced. She had the softest voice and gentleness that reached past my defenses. But she also carried a firm, steady presence, filled with conviction.

As we began to talk more, we grew closer. There was something about her that felt like peace. Like healing. Like home. Whether it was the way she calmed me when I opened up, or the way she didn't judge me when I told her what had happened in Iraq—her response was always the same: GRACE. Coming from a military family herself, I wondered if she would understand. And she did. She once told me that when she'd spotted me working from across the store, it was like a dark cloud hung over me. She was right. I was far from who I used to be—removed from my element, lost, aimless. I thought my life was over.

I remembered when I finally opened up to her about what happened that day in Iraq. My hands were trembling. My voice, uncertain. I did not know what to expect, maybe. Pity. Or worst of all, silence.

"I need to tell you something... something I haven't really spoken about in a long time."

(She nodded, saying nothing, just waiting, giving me the space.)

"That night in Iraq... the accident. It was all my fault. My mistake. And it cost a good man his life. I carry it every day. Every hour. Some nights, I still hear the shot.

"That must be unbearable," she muttered.

"It is. That's the thing—I can't fix it. I can't undo it. And I don't know how anyone could love a man who did what I did." She didn't flinch. Didn't look away. Just reached out and took my shaking hand into hers.

"Will, I don't love you because you're perfect. I love you because even in your pain, you're still trying. I see your heart. I see the weight you carry.

"But through her compassion, her words, and her spirit, she showed me something different.

"I thought if I told you, you'd leave," I uttered, lowering my head.

"You think your past makes you unlovable. But it's your honesty, your brokenness, that makes you real. I don't see a man beyond saving—I see a man that grace hasn't given up on. And neither will I," she stated.

God's grace is everlasting. It doesn't run dry when we stumble. It doesn't expire when we've fallen too far. I've learned—sometimes the hard way—that grace isn't a one-time pardon; it's a constant presence. A covering. A quiet mercy that stays with us when the rest of the world walks away. I didn't earn it. I still don't understand why I was given another chance. But I knew at that moment that God's grace doesn't leave us where it finds us—it lifts us, restores us, and walks with us until we can stand again.

There are no words strong enough to capture what Priscilla has meant to me. Her quiet strength, unwavering love, and deep well of grace have carried me through some of my darkest seasons. She married a man burdened by grief, guilt, and brokenness—and chose to stay when many would have walked away. Where I saw failure, she saw someone worth fighting for. Where I questioned my worth, she offered dignity. And when I struggled to believe in God's mercy, she embodied it.

Through long silences, recurring nightmares that I was not willing to share, and seasons of emotional distance, Priscilla remained. Not passively, but with fierce tenderness—speaking truth when I needed correction, extending kindness when I deserved none, and holding space when I had no words at all. She has been my anchor in the storm, my mirror when I've lost sight of who I am, and the heartbeat of every second chance I've been given.

Her love has been both shelter and compass, helping me find my way back not only to myself, but to purpose, to healing, and to hope. This journey may be mine to tell, but it was never one I walked alone. Priscilla's quiet courage is there—an enduring gift from God—lifting me with grace, loving me with conviction, and believing in the man He is still shaping me to be.

Priscilla is the living, breathing representation of God's grace in my life. Not the kind of grace you read about in abstract terms, but the kind that shows up with a steady hand and a soft voice when you're drowning in guilt. She didn't just love me—she redeemed something in me. Through her patience, her faith, and her unwavering presence, she embodied what I had read about but never genuinely believed was meant for me: that God's grace could still reach a man like me. That I wasn't too far gone. That love could be real again. In every moment she chose to stay, to listen, to believe—I saw God's mercy made tangible.

20

THE CONSEQUENCES OF FAILED LEADERSHIP

Philippians 2:3–4 - King James Version (KJV 1900):

"Let nothing be done through strife or vainglory; but in lowliness of mind let each esteem other better than themselves. Look not every man on his own things, but every man also on the things of others."

It has been twenty years since the events of June 23, 2005, two full decades marked by reflection, sorrow, and a burden that never lifted. The passage of time has dulled many memories, but not this one. The pain, the guilt, and the profound mental unrest that I caused have never truly gone away. They live with me. They haunt me. Even now, all these years later, I still wake from nightmares, sometimes afraid to close my eyes. There are nights I avoid sleeping altogether. And yes, there are still nights when I wake in a cold sweat, humiliated by something as simple yet devastating as a wet bed, the result of some memory I can't escape, relived in my dreams.

I have been to therapy more than once. I have sought help in chaplains' offices, in counselors' rooms, and in quiet conversations with those I hoped might understand. I have tried to make peace with myself. I have been married twice, and though both women saw strength in me, I know they also saw the cracks, the parts of me shaped and sometimes fractured by what happened that day. No amount of time, love, or healing has managed to uproot the sorrow or silence the guilt that took root on that June morning in Iraq.

Sometimes I wonder whether that day, what happened, what failed to happen, should be taught in leadership courses throughout the military. Perhaps every cadet and midshipman, at every service academy and ROTC program in this country, should be required to study it. Not as a footnote. Not as an afterthought. But as a case study in what happens when leadership fails.

Leadership has many definitions. Scholars, generals, and presidents have filled libraries with their theories of what makes a great leader: courage, vision, character, and compassion. Others write about good leadership: competence, fairness, and the ability to inspire. But few truly explore the cost of bad leadership, not merely ineffective or unpopular, but leadership that is negligent, toxic, or unfit for the moment. That kind of leadership can have catastrophic consequences. And in war, it can kill.

What happened on June 23, 2005, is one such story. It is not one of heroism or valor. It is not a tale of triumph against odds. It is a cautionary tale, one where poor decisions, delayed actions, and misjudgments collided in a way that changed lives forever. Including mine.

This chapter, this entire memoir, is not written out of vanity or vindication. It is written out of duty. A duty to those who were there, a duty to those who paid the ultimate price, and a duty to those who may someday find themselves in a moment of consequence, not unlike mine.

And so, with humility and conviction, I offer my life experience to the United States Army. It is my deepest hope and prayer that this work might one day serve as required reading for all future Army officers. Let it stand as a mirror, a warning, and a lesson.

To those who will lead soldiers in the years to come: never forget the sacred weight of the responsibility you carry. Your actions, your decisions, and even your smallest habits can mean the difference between life and death, trust and betrayal, honor and shame.

I challenge you to lead with courage, humility, and vigilance. May you learn from my failure so that you never have to endure the pain of living with one. And may you honor the lives entrusted to your command—not only through words, but through unwavering discipline and integrity.

Because leadership is not just about guiding troops through battle. It is about bearing the weight of every life entrusted to your care. And when that leadership fails, the consequences are not abstract, they are written in blood, grief, and sleepless nights. Officers don't just wear rank; they bear the burden it brings. They excel under pressure, and every decision and action taken by every person under their command is their sole responsibility.

Show leadership, live standards, and doom complacency.

Learn what I had to learn the hard way.

EPILOGUE:
A SOLDIER'S
REFLECTION

A SOLDIER'S LAMENT

Letter to the Family of Sergeant Joseph M. Tackett

To the parents, family, and friends of Sergeant Joseph M. Tackett,

There are no words to fully capture the depth of my sorrow for the loss of your son. Saying "I'm sorry" feels utterly inadequate when my negligence forever altered and ended Joseph's life. Even now, nearly twenty years later, I struggle to believe it happened. His death haunts me like a recurring nightmare, and some days I cannot summon the strength to get out of bed. Other days, I feel unworthy of any small measure of joy, as though doing so would betray Joseph's memory. This is why it has taken me so long to reach out: I cannot fathom the grief I've caused you.

I recognize that a simple apology cannot erase what occurred, nor do I expect your forgiveness. But I pray you will understand this one promise: your son's life will not have been lost in vain.

Though I did not truly know Joseph well personally, everything I observed about him spoke to his character. He was a dedicated soldier whose good humor lifted everyone around him. He led by example, and I witnessed firsthand his potential for leadership. No one ever spoke ill of him. He was, without question, an exceptional soldier and a remarkable young man.

Mr. and Mrs. Tackett, I believed myself to be a man of honor, and I believe Joseph was, too. I apologize for how long it has taken me to extend this apology. For years, I wrestled with how best to approach you. Today, I ask that, if you can, you find it in your heart to forgive me. If forgiveness is not possible, I understand completely; I failed in my duty to protect your son and to uphold the responsibilities of my commission.

In my own struggle with guilt and despair, I turned to Scripture and prayer. I discovered that while the Bible acknowledges deep suffering, fear, guilt, and heartache, it nowhere declares that any sin, even one as painful as suicide, is beyond God's mercy. I learned that true healing begins when we claim God's strength and accept the support He provides through others.

I do not presume having the right to ask your forgiveness. Yet I do ask it, humbly and sincerely, in the hope that it might bring some measure of peace to us all.

With profound respect and deepest regret,

Willie Davis, Jr.

STILL ON DUTY

It is my duty to write this memoir because silence only compounds tragedy. If I were to keep this story buried, if I allowed the weight of guilt, shame, and fear to muffle the truth, then the lessons born of pain would be lost. Leaders must speak not only from the podium of success, but from the valleys of failure. It is in those valleys, when everything crumbles, that the most valuable insights emerge. My failure cost a man his life. To hide from that would be to betray his memory all over again. This memoir is not for vindication; it is for illumination.

By laying bare my mistakes, I offer others a mirror, not to shame them, but to awaken them. It is easy to believe that leadership is about strength and confidence alone, but true leadership demands accountability and reflection. I was entrusted with the lives of my soldiers. I failed them. Not because I lacked courage, but because I lacked the vigilance and maturity required in every moment. If one officer reads my story and tightens their grip on responsibility, then perhaps another tragedy can be averted.

There are moments in a leader's journey where their actions ripple far beyond what they can see. The ripple of my negligence extended far—touching Sergeant Tackett's family, my comrades, and those I love most. I am haunted not only by what I did, but by what I failed to do. Writing this memoir forces me to confront each ripple. It holds me accountable to the past and charges me with the responsibility of using that past to prevent future suffering.

I owe this memoir to every young lieutenant stepping into command for the first time. They deserve more than checklists and regulations; they deserve lived truth. They deserve to understand what can go wrong when discipline becomes a habit, and habit becomes complacency. They need to hear the cries beneath the courtroom silence, the ache of a broken marriage, and the endless replay of a single moment gone terribly wrong. My story is not a cautionary tale, it is a warning siren, one I pray they never ignore.

It is also my duty to write this for those who have fallen, not only Sergeant Tackett, but the many leaders who have crumbled under the unseen weight of guilt. We do not often speak of the psychological aftermath, of how shame can eclipse every joy, how the burden of remorse becomes a second prison. In writing this, I offer a lifeline to others drowning in their own private despair. I want them to know they are not alone, and that healing, though never complete, is possible.

This memoir is for Sergeant Joseph Tackett. He did not ask for his name to be remembered in this way. He did not deserve to be a chapter in another man's regret. But his life mattered, and my accountability to him does not end with the sentence I served. It continues in how I live, how I remember, and how I help others, never to forget the cost of carelessness. To remain silent would be to let his memory vanish beneath paperwork and time. I refuse to let that happen.

Lastly, I write because I must. This is my repentance, my reflection, and perhaps, my redemption. I am not seeking pity. I am seeking purpose, something to rise from the ashes of a life burned by one tragic moment. Writing is the only way I know to offer something meaningful in exchange for all that has been lost. If this memoir allows even one leader to lead with more care, one soldier to speak up, or one heart to soften with understanding, then it will have been worth every painful word.

TO MY BROTHERS AND SISTERS IN ARMS

There's a reason so many of us don't make it home whole. And I'm not just talking about visible wounds. The truth is, for many who have served, the battlefield doesn't end when the war does. We carry it with us into our homes, our marriages, our memories, our sleepless nights. Some of us carry guilt. Some carry grief. Some carry silence so loud it drowns out everything else.

In 2022, more than 6,400 veterans died by suicide—nearly 17 every day. That's not just a number. That's a roomful of names. Brothers. Sisters. Warriors. People who gave their all, then came home and felt like they no longer belonged. People who believed, deep down, that they were beyond saving.

I was one of them.

I've stood at the edge of that darkness. I've asked the questions no one wants to say out loud. I've looked in the mirror and not recognized the man staring back. I've held on to shame so tightly it nearly broke me. But somehow, in the lowest of moments, grace found its way in—not through some grand, sweeping act, but through the small, faithful kindness of others.

A chaplain handed me a Bible I didn't ask for. A friend sat with me in silence when I had no words. A woman loved me through my worst and believed I was still worth saving. Those moments didn't erase the pain. But they cracked the door open. And through that sliver, I crawled my way back.

I won't lie to you; healing doesn't come quickly. It doesn't come easy. But it does come. And more importantly, it's possible.

So, if you're reading this right now and you feel like you've fallen too far, please hear me: you haven't. There is no uniform that can protect you from human pain, but there is also no failure so deep, no guilt so heavy, no regret so sharp that it places you beyond the reach of grace.

You are not alone. You never were. And your story— it's not over yet. Stay. Fight. Breathe.

We still need you here.

— *Willie Davis, Jr.*

A LETTER ON COMMAND AND CONSEQUENCE

To the Future Cadet and Midshipman. You may be reading this as a high school student dreaming of earning your commission. Or perhaps you're already in ROTC or attending a service academy—studying hard, memorizing ranks, and preparing for the challenges ahead. Whether you're bound for the Army, Navy, Air Force, Marine Corps, Coast Guard, or Space Force, I want to speak to the leader you are becoming.

I remember being you.

I remember what it felt like to put on the uniform for the first time, to feel purpose in my stride, pride in my chest, and a future full of promise. I believed I could lead because I had studied leadership. I memorized doctrine, passed my evaluations, and checked all the boxes. But leadership, *real leadership*, is not about tests or inspections. It's not about how loud you call cadence or how sharp your salute looks. Real leadership shows up when things fall apart.

Military leadership is not about perfection. It is about responsibility. It is about owning your decisions, *all of them.* It's about standing between danger and those who follow you. It's about having the courage to listen, to speak the truth, and to make decisions when no one else wants to. Most of all, it's about carrying the burden when others can't.

I made a mistake, one that changed lives forever. I was negligent, and no explanation could erase that truth. It was my responsibility, and I failed. And I paid a heavy price. Not just in uniform or status, but in guilt, grief, and the long road toward forgiveness.

Leadership does not end at failure. It begins again in the act of taking ownership.

So, to you, future cadet or midshipman: prepare yourself with excellence. Train hard. Know your craft. But never forget that the most important part of leadership isn't in your rank or ribbons, it's in your heart. Lead with humility. Lead with empathy. Listen to your NCOs and petty officers. Protect the people entrusted to you. You will carry their lives in your decisions. Don't ever take that lightly.

And if the day comes when you fail, and it might, do not let it define you. Let it refine you. Because true leadership is not about how well you command in comfort, but how courageously you lead through crisis.

Earn your commission not for glory, but for service. And when you lead, lead them well, they deserve nothing less.

- Willie Davis Jr.

ACKNOWLEDGEMENTS

Psalm 91:11-12:

For He shall give His angels charge over you, to keep you in all your ways. In their hands they shall bear you up, lest you dash your foot against a stone.

Life has its twists and turns. And during my time of trial and tribulation, I have been blessed to have a very loving family and some great friends who I believe were placed in my life to help me come to terms with what I have been dealing with. Some of you have been in my life for years, while only a few have been in my life for just a few months or a few years, during which I was learning how to live with causing SPC Tackett's death. I am incredibly grateful that God put these people in my life. People need people to be by their side in truth in any ordeal. I am and was blessed by their presence and will forever be grateful to them for their love and support.

To my friends, **LTC (R) Danielle Smith and Shannell Marie**:

First and foremost, thank you for standing by my side when I needed you most. I am profoundly honored and proud to call you both my friends, especially during those harrowing days in Iraq. Your selfless service to our country, alongside your unwavering personal support, means more to me than words can express.

Thank you for offering me a shoulder to cry on, a listening ear when I felt isolated, and constant encouragement when hope seemed distant. In moments when the weight of loss and guilt threatened to overwhelm me, you both stepped in, ready to listen, ready to console, and ready to lift me up. I will never forget your compassion, your strength, and your belief in me. For all you have done, and all you continue to do, I am eternally grateful.

To my mother, **Hancey Davis**, and my late father, **Willie Sr.**, whose love and guidance shaped me.

I am blessed and highly favored by God with you as my parents. I hope that I make you proud. This was a great tragedy for an untold number of people. But with both of you by my side in God's love and in faith, I am highly blessed.

To my former Battery Commander, **Brigadier General Alric Francis:**

Sir, you were the very embodiment of what it means to be a soldier's officer. Your leadership, integrity, and commitment to those under your command left an indelible mark on me. I looked up to you with great admiration, and your example set a high standard, one I aspired to meet, even if I did not always succeed.

For the moments when I fell short, I offer my sincere apology. Yet I remain grateful for the example you set, the guidance you provided, and the inspiration you gave. Thank you for your leadership, your service, and the lasting impact you had on my life and career.

To Colonel Daniel A. Pinnell:

Thank you for standing by my side through the most heartbreaking chapters of my life and career. Your unwavering support, sage counsel, and steadfast leadership sustained me when I needed it most. You, and alongside you, Brigadier General Alric Francis served as beacons of integrity and strength, modeling the officer and mentor I aspired to become. I am deeply grateful for every conversation, every piece of advice, and every moment you invested in guiding me through that tragic ordeal. Your example, your friendship, and your faith in me have left an indelible mark on my heart.

To my therapist, **Mr. Rick Ennis**, Louisville, Kentucky:

Thank you for guiding me through the darkest chapters of my life with kindness, wisdom, and unwavering support. Your compassionate ear and sage advice have carried me from a place where I once contemplated ending my own life to a place where I wake each day with renewed purpose and resolve. You taught me how to face my guilt, harness my grief, and transform my pain into action.

I am especially grateful for your help in shaping this book, teaching me how to speak honestly from the heart, to extend a sincere apology to the Tackett family, and to share my story in a way that honors both my own journey and their profound loss. Your insight and encouragement have been essential, and I cannot thank you enough.

To **Kory M. Saunders** and **Kya Saunders**—sisters not by blood, but by bond.

Thank you, eternally, for being not only the best sisters a brother could ever hope for, but also the truest of friends during the darkest chapter of my life. Your unwavering love and support, especially through your letters and uplifting messages while I was in confinement, became a lifeline for me, steady, consistent, and deeply healing. In a place where hope was scarce and peace of mind even scarcer, your words brought light, comfort, and the sense that I had not been forgotten.

Kory, I am especially grateful to you. Through your letters, you allowed me a window into your world, a glimpse into the progress, growth, and joys of your life, even as I was stalled in my own. Those words gave me something to reflect on, something to hold onto, and more importantly, something to believe in. They reminded me that life was still moving forward and that I still had a place in it.

From the bottom of my heart, thank you. You helped carry me through.

To my lovely wife, My Queen, **Priscilla**:

Thank you for choosing to marry a man who, at times, could barely look at his own reflection. A man burdened by grief, guilt, and the weight of a past he could not undo.

You loved me when I struggled to love myself. You saw light in places I thought were consumed by darkness. Through sleepless nights, restless silences, and the pain I often couldn't put into words, you remained. You held my hand when I was drowning in shame, and reminded me again, that brokenness is not the end of the story.

Your compassion became my safe place. Your strength became my stability.

Your belief in me, when I had lost faith in everything, including God, became the very evidence of His mercy.

You have been the quiet miracle in my life, the living proof that grace doesn't always shout. Sometimes, it simply shows up, every day, with patience, forgiveness, and a love that refuses to give up.

I am endlessly grateful to God for you.

Not just for your love, but for the way you reflect His: unconditional, healing, and transformative.

In His wisdom, He gave me not what I deserved, but what I needed, and what I needed was you.

Finally, I give my deepest thanks to **God**, whose boundless love and mercy have carried me through life's most painful trials. It is by His grace that I have been granted the courage and the opportunity to write this memoir and share my story. My prayer is that, within these pages, someone who feels broken will find the strength to hold on, to continue doing God's will, and to use their own experiences to uplift others. I remain imperfect and often struggle with the weight of my past, but through faith, I press on, hopeful that this work can be a beacon of light for anyone in need of comfort and encouragement.

In Honor of
SGT Joseph M. Tackett
(1982–2005)

Whose life, laughter, and leadership left a mark that can
never be erased.
This memoir is a lament, a confession, and a promise—that
his memory will never be forgotten.